IN INDIAN TERRITORY

IN INDIAN TERRITORY

A STORY OF TWO FAMILIES

BOB JACKSON

Copyright © 2021 by Bob Jackson

All rights reserved.

No part of this book may be reproduced in any form or by any electronic or mechanical means, including information storage and retrieval systems, without written permission from the author, except for the use of brief quotations in a book review.

❦ Created with Vellum

For my wife

1
LEON, IOWA

Lew was born in 1875 to Caleb and Molly Riggins. He was born with a keen power of silent observation and with an analytical mind.

He was not endowed with a love of people. Rather, he had focus—on a project or a larger goal. He was happy to spend time with people he worked with, but aside from work, he didn't have time for people; he was always too busy. He was all business.

He grew up in a farmhouse a quarter mile from Leon, Iowa, population 900. It was the county seat. It had a courthouse, a small professional building housing two lawyers, a part-time doctor, and a dentist; a wooden church, a bank, a small hotel, a one-room school, a general store, a farmers' equipment and feed store, a small library, and some residences.

Lew had ten siblings. (He would have had 12, but two had died in infancy.) The house was crowded, and the dining-room table couldn't seat the whole family. Latecomers had to stand and juggle their dishes.

There were seven boys and three girls. Four of the boys were older than Lew; two were younger.

As soon as a child was big enough to hold a hoe, he —or she —was expected to help with the farming. It was hard and tedious work, especially when the summer sun made it unpleasant to be outdoors, nonetheless working on a farm. And their father expected long hours.

The family got by, but just barely.

Lew did odd jobs in the winter to accumulate some cash. Most were in Leon. A lot of his work was interior painting. He developed a reputation of being reliable and doing good work.

Each boy in the family had two sets of work clothes. When one got dirty, he switched to the other and their mother washed the soiled one. Each boy also had a set of clothes for church.

All of Lew's clothes were hand-me-downs. A casual observer could see that they had been patched and mended several times. He had two bib overalls, two shirts to go with them, and a pair of pants plus a shirt for church.

Someday I'm going to buy all my clothes new, in a store.

He and his siblings walked to and from Leon every weekday, to attend the school. Even though going to town, they wore their work clothes, which embarrassed Lew.

People must think that we're a gang of ragamuffins.

My only pleasure is chess. But, except during the winter, there's no time for it.

Lew and his father played, when his father could take time away from the other children. Lew picked up on it immediately and after a while was formidable. He beat his father as often as his father beat him.

It was a religious home. The family said grace before every meal. They walked to the early church service every Sunday morning. Back home, everybody put away their Sunday best clothes, changed into work clothing, and got back to farming.

He grew up in a crowded farmhouse near a small town in southeast Iowa. He had ten siblings. (He would have had twelve, but two had died in infancy.) He didn't need Malthusian theory to know that overpopulation results in scarcity.

It was a religious home. The family went to church every Sunday. And they said Grace before every meal.

What I like best about church is the music. The church can't afford an organ, but it has a nice piano. And the congregation does an acceptable job of singing hymns.

My favorite is "The Battle Hymn of the Republic." I like it because people sing it with more gusto than with other hymns. And the words of the second line do something for me. I'm not sure what. Maybe it's like a call to action: "He is trampling out the vintage where the grapes of wrath are stored."

Lew developed an opinion about farming early on.

Farming's not for me. It's monotonous and boring. We work long hours, often in intense heat. And despite our hard work, we're poor, on the edge of poverty. We barely have enough to eat, and we can't afford to buy anything nice or do something fun.

From my observations, and from what my teachers hasve told me, lawyers, doctors, bankers, and most other educated people, work indoors, often in offices, and appear to do better than others. well.

It looks like education is my way off the farm.

So, he focused all his energies on school, which he found he really liked. He learned quickly and effortlessly, and he did well.

The teacher in Leon this small, rural was school were starved for eager students. She liked Lew and made no efforts to hide that. This might have resulted in problems with the other children, but it didn't because they liked him. He made friends easily and knew everyone in the school.

The teacher responded to Lew''s attitude and achievements by going out of their way to help him go beyond the other students in his classroom subjects. She One teacher in took particular charge of him and took charge of Lew. encouraged him.She frequently commented, "You'll do well in life."

LEW CONTINUED to think about the future.

I want to decide what I'm going to do after I finish school. I think I"ll talk to interview some people to see what different jobs are like.

He took it upon himself to interview the town's two lawyers, its banker, and its doctor. He talked with the also interviewed two shopkeepers about business matters. After ruminating, he made his choice.

I want to be a lawyer. A lawyer's his own boss. He does exciting things like meeting with clients and trying lawsuits. And he makes a comfortable living. So, that's what I'm aiming at.

When Lew was 17, the teacher at the high school said, encouraged him to go further. "I've done about as much as I can for you. If you want to practice law, you need to go to college, so you can make your way in the world. Don't waste your talents."

That night, Lew spoke to his father. "I want to go to college in Des Moines. I hope that's okay. I won't be here to help with the farming."

"Why do you want to go to college? No one else in the family has gone to college. And we all seem to get along without it."

"I don't want to hurt any feelings, and this is not personal. It's just that I don't want to stay in farming. I don't like it, and I want to do something else. I've thought about this for a long time. It's not a spur of the moment impulse."

"And what do you want to do other than farming?"

"I want to be an attorney."

"You're setting your sights high. And I can see that what you want requires college. Let me think about it. And I want to talk to your mother about it."

Because some of the children were sitting in the same room as Lew and their father, they overheard the conversation. No one said anything about it that night.

The next day, when they were working in the field, the oldest boy, John, confronted Lew.

"You think you're too good for farming? Are you better than the rest of us? The family has been farming this land for a long time. No one else has tried to quit it."

"No. I'm not any better than the rest of you. It's just that I'm different. I want to work with my head, not with my hands."

"What about the effect on us? We can barely get enough done with your help. What are we supposed to do if you leave?"

"I've thought about that, and I'm afraid I don't have a good solution. Possibly, Father can find occasional help when it's most needed. The extra harvest should cover the wages."

"That's just a makeweight. You're lazy and don't want to work hard. You want to quit even if it hurts the family. It's selfish."

"I can understand your anger, although I don't think it's warranted. If Father agrees to my leaving, I'll still be part of the family, although I'll be physically distant. I hope we can all pull together and adapt to the change."

"That's easy for you to say. But some of us hope that Father says no."

Their father never took long to make a decision. Lew and his brothers anxiously awaited it.

"Your mother and I want you to do what's important to you. You've worked hard as a student, done well with your studies. We don't want to hold you back."

The older boys were glaring at Lew. He could see that they

were disappointed and angry. But he was not willing to give up on college.

"And it'll work out with you being gone. One less worker; one less mouth to feed. The one balances out the other. But we can't support you. I think you know that."

I've saved enough from odd jobs in the winter to get myself to Des Moines and start to get started there. And I figure I can find some work to support myself."

It'll be a tough row to hoe, but you'll make it. You've developed the determination to do whatever you set your sights on.""

A FEW WEEKS LATER, Lew he said his "goodbyes." It was a Sunday, and the family had just returned from church. It assembled to see him off.

He didn't change his clothes. He wanted to arrive in Des Moines in his Sunday best.

His mother was tearful. He hugged her.

"Mother, please don''t worry about me. I'll be all right. And I'll write to let you know how I'm doing." And I'll return for monthly visits."

Lew felt terrible that his leaving caused his mother to cry. Without making a conscious decision to lie, he did so. He knew deep down that he would not frequently return. He just wanted to appease her.

He embraced his father and said, "Thanks for teaching me how to get things done with persistence and hard work."

"I appreciate that, son. And as you make your way through life, keep in mind these words from Proverbs 22: "Train a child in the way he should go, and when he is old, he will not turn from it'."

"I'll do that. Thank you for being understanding about my

leaving. I feel that I have your blessing to do what I think is right for me."

He extended his arm and shook hands with each brother. But the ones older than he would not look him in the eye. He hugged each sister.

To avoid awkwardness, he said something to each of them. "Keep an eye on the place." "Help Mother with her chores." "Keep up your good work." "Watch out for the little ones." "Help Father with his heavy work." But these suggestions just elicited more angry glares from the older boys.

To the whole family he said, "I'm sad to be leaving home, but I know I'm doing the right thing. I'll miss you. But I'll return for visits."

And to himself he thought, He thought, *I don't want an excess of emotion.*

So, he quickly left the house and got into a buckboard driven by a neighbor.

It took him more than an hour to relax. He kept thinking about the ugliness in connection with his leaving.

Maybe one day I'll be able to send money.

This helped him let go of the ugliness.

I feel like a yoke has been lifted from my shoulders. A weight has been removed. Home life was choking me.

2

COLLEGE

When Lew got to Des Moines, he located the school and found a nearby place to stay.

The next day, he went to a small clothing store frequented by students. He got two pairs of pants and two shirts. He left the store wearing one of each.

He wrapped up his old clothes and shipped them to his mother.

The next day, he went to the school and asked the first student he encountered how to get started. The student directed him to the registrar's office. He went there and enrolled in Drake University. He negotiated a payment package under which he would pay the fees for the first term, with interest, at the end of the term.

As part of the process, he met with a faculty advisor.

"What courses do you want to take?"

"I'm not sure. I'd appreciate your guidance on that."

"What do you want to do with your life?"

"I want to be an attorney at law."

"Why?"

"Because I want to work indoors, and I want an opportunity to earn a decent living. I want to live by my wits, not by my hands."

The advisor thought about that. "Based on that aspiration, I suggest the classics: philosophy, history, literature, art. That should provide a solid foundation for becoming an attorney at law."

Lew accepted the recommendation and selected his initial courses.

"That's a heavy load of courses. It's the maximum that we allow. You could take fewer courses and take some of the heat off yourself."

"No, thanks. I want to finish at Drake as soon as I can."

He clerked in a campus store that handled a variety of goods, including clothing, books, and drugstore items. He excelled at his job. It was not long before the owner made him the manager. In that capacity, he took inventory, hired and trained clerks, and ordered fresh inventory. He got frequent pay increases.

He worked nights and weekends to finance his tuition and living expenses.

When he could, he played chess. There was no formal chess club, but people who wanted to play could always find each other. And it was only a matter of time before he was regarded as the person to beat.

In addition to his studies, Lew joined Drake's debating society.

I joined to see what it's like. But I'm finding that debate comes easily to me.

He excelled at it, and his reputation spread rapidly. People soon avoided opposing him.

A lively debate involved The Indian Removal Act of 1830. President Jackson had pushed for the Act. He wanted white farmers in Georgia and Alabama to have the fertile farmland of the Indians.

I think that the Act is immoral. The federal government has no business mistreating Indians.

He argued that the mistreatment of the Indians far outweighed any benefit to the nation. But he lost the debate.

Lew didn't stir after the moderator announced the decision. It was as if he were frozen.

Damn. That hurts! I'm used to winning. I don't understand why I lost this one. I worked hard, preparing my argument. It felt like I did a good job of presenting it. What went wrong?

He spoke to one of the judges.

"Mr. Raven, I thought I did a good job. Would you care to give me your thoughts on why I lost?"

"Sure, Lew. You got too involved emotionally. You left no doubt about how you feel about the Act. But an argument should be objective, not subjective. The best argument is a good explanation, devoid of emotion."

"Thanks, Mr. Raven. I appreciate your candor. I've learned something valuable."

Lew never forgot this lesson. He was always vigilant not to become personally involved in a position or argument.

NOTWITHSTANDING THAT HE was younger than most of the students, Lew got higher than average marks. His teachers were impressed with him. But he was not happy.

As Father had predicted, it's a hard life. Between my studies, debating, and work, I have little time for anything else. I suppose I should visit my family, but my schedule never seems to permit it.

He did manage a visit during the Christmas season. But that

was an unusual opportunity. There were no classes and most of the students were gone. So, the store was closed.

Home was home. Nothing had changed. His older brothers were still furious about his leaving. His mother wanted him to give up on college and not return to Des Moines. His father was fixated on the family's attending the church's Christmas service, putting on a good front for outsiders to observe.

All in all, the visit was hard on Lew. He felt anxious most of the time. He could hardly wait to return to Drake.

HE WAS an enigma to his classmates. They couldn't understand his single-minded devotion to his studies and how he could get his marks without grinding. But they liked him. "Hey, Lew, how about grabbing a beer?"

He sometimes accepted, not because he wanted to be with them, but because he viewed it as a professional opportunity, a good investment of his time.

These guys are local. They'll probably succeed—find good jobs or start businesses. They'll need attorneys. Getting to know them could be useful someday.

A minor part of his decision was the chance to have a beer or two. Lew had found that he liked drinking. He never did it just for itself. He didn't have time for that. But if it was part of an activity that he deemed worthwhile, it was a nice side benefit.

Before graduating, Lew visited a tailor and ordered a dark, three-piece suit. He had observed that this was the "uniform" for an attorney in Des Moines.

3
LAW PRACTICE

Now that I've graduated from college, I want to "read law." I want to become an apprentice with an established lawyer who will create and supervise a course of clerking and study designed to teach me the law.

He interviewed three people. One commented on his credentials. "Your professors have highly recommended you. One said that you have developed an extraordinary ability to concentrate intensely on what's important. Your mathematics professor said that when you set your mind to something, you become very determined."

All three attorneys offered Lew a job, and all three offered more compensation than was typical for apprentices.

I'm not going to base my decision on compensation. I want to work with the attorney with the most diverse practice.

READING law was less demanding than the combination of college, debating, and the job he had had before. At last, he visited his family more often and socialized a lot. He talked to people easily. He became good at small talk. It just took practice. He liked parties and, as a result, he was invited to a lot of them. People soon regarded him as a particularly eligible bachelor.

It's worth my time to attend these parties. The people I meet are successful. Most are businessmen and professionals. They are socially prominent. No telling how they could help advance my career.

Plus, the hosts serve liquor, which helps me relax.

One evening the hostess of a party introduced him to Pearl. She had moved to Des Moines from Ohio with her father. He had wanted to start a clothing business, but he died from a respiratory illness shortly after they had arrived. She knew shorthand and typing and, having been thrust into the role of supporting herself, was working as a legal secretary and living in a boarding house.

She was pretty, well spoken, and dressed fashionably. Her father had seen to that.

I'm attracted to her. I wonder if she could become attracted to me?

He pursued a courtship. They were both gregarious and enjoyed parties, to which hostesses began inviting them as a couple. They took long walks together. They played parlor games with other couples. They participated jointly in political activities.

As they got to know each other better, they became more intimate about what they shared.

I've developed strong feelings toward her. I'm going to summon up my courage and propose marriage.

On a Saturday morning of a beautiful spring day, Lew was walking, more deliberately than usual, towards Pearl's boarding house. He knocked on the door and asked for Pearl. The proprietress recognized him and went upstairs to fetch her. When she arrived, Lew signaled for them to go into what the proprietress called her parlor. He closed the doors behind them. Pearl felt alarmed; gentlemen didn't do that.

"I have something I want to discuss with you," he said.

Pearl maintained a long silence. "What is it?"

"I'm having trouble finding the right words. Ironic, really. Words always come easily to me. But today they fail me."

"Why don't you start talking, and maybe what you need will come to you."

"Good idea. I want you to know that I really enjoy being with you. And I admire your skills, your resourcefulness, and your judgement. And we share the same values."

"Thank you. I appreciate those words. But I doubt you came here to tell me those things."

"You're right, as usual. My real message is—"

Pause. Then hurrying. "I don't often get excited about things. But I've been excited about you for quite a while. I love you, and I doubt that anyone could love you more than I. I want to marry you and be with you always."

Pearl, playing with him. "Isn't there a question you should be asking me?"

Lew, taking it seriously. "A question? A question? Oh! Will you marry me?"

"I've been hoping for this. I love you too. Of course, I will."

Lew finally relaxed and they embraced and kissed.

He paused to catch his breath and to think about what had just transpired.

I don't want to rush things.

"I hope you're willing to wait until I pass the bar, so I'll know I can support you."

"How do you know you'll pass the bar? And if you do, how do you know you'll succeed in a law practice?"

"I've always done what I set out to do. And I've been working hard at my apprenticeship. My mentor is helpful."

After a moment's thought Pearl said, "I have confidence in you and I think your plan is sensible."

When his mentor declared him ready, Lew sat for the bar examination. He was notified by telegraph that he had passed. He was so excited that he couldn't sit still. He left his mentor's office without telling him the news and hustled to the office where Pearl was working. "Pearl! Set the date and make the arrangements. I passed the bar! I'm a happy man."

His excitement was contagious.

"Wonderful! I can't wait!"

Lew lowered his guard. They embraced and kissed, which they had never before done in public.

LEW WAS ADMITTED to the Iowa bar in 1900.

I'm going to practice by myself. I want to be my own boss. And I don't want to pay other people's overhead.

His mentor, usually reserved and not at all demonstrative, smiled and put his hand on Lew's shoulder.

"You're a natural. You were born to be a lawyer. You listen to people and absorb what they say. And you're a good judge of people. I expect great things from you. Good luck!"

He opened his office in the city of Des Moines. Pearl did his secretarial and clerical work and still had time to work for others; she was saving up money. His practice started slowly, but increased in a matter of months. He was known in the right circles, and he acted as if he cared about people and their problems. Understanding and analysis came naturally to him. He didn't have to work at them. He did mostly real estate and trial work, was clever at them, and the results were good.

WITH HIS PRACTICE UNDERWAY, he and Pearl married. They decided against having the ceremony in his hometown. They wanted it to be in Des Moines, because they had many friends there. Most of his family attended, even though it meant taking time off from school and the farm. They loved Lew and were anxious to meet Pearl.

The wedding went well. The city guests didn't appear to pay any attention to the fact that a farm family was in attendance.

The beginning of the reception was a bit awkward. Lew's family neither looked like nor acted like the other guests. They felt self-conscious and kept to themselves in a corner of the room. Pearl noticed this and took them one at a time to meet some of the other guests. The rest of the family relaxed and, on their own, mingled with the crowd.

The other guests observed this. They saw it as a show of Pearl's affection for Lew.

After the wedding, Lew and Pearl moved into an apartment that she had found and rented for them.

4

LAND OF OPPORTUNITY

Lew and Pearl made enough money to live comfortably, but he became restless.

I want to do better financially. Lawyers prosper in an environment of rapid change. And Iowa is relatively stable. I want to look for alternatives.

He had heard about Indian Territory in what would become part of Oklahoma. He read what he could find about it. He learned about the resettlement of Indians there. He learned that whites were moving in and life there was changing rapidly.

Indian Territory might be a better opportunity for me than Iowa.

"Pearl, what would you think of resettling?"

"Where?"

"I'm thinking of Indian Territory."

"Don't say that."

"The population there is growing. Things are changing. The Government distributed Indian lands to tribal members, each receiving 160 acres. They have to leave the communal lands where they had lived and relocate to their own land. Whites are buying land to farm and ranch and starting businesses. Those things create work for attorneys. There are probably not a lot of attorneys there yet. I think we would find more opportunity there than here."

"But our friends are here."

"I'm not happy about that. But think of it! It's largely unsettled. It's a frontier, new and exciting! And we'll be a part of it, in from the start! No telling what we'll accomplish."

"Let it sit for a few days and see if you still feel the same way."

At dinner, a few days later, she raised the subject.

"Are you still thinking we should move?"

"Absolutely. I'm ready to do it as soon as we can."

"I've trusted your judgment this far. I might as well trust it further."

Lew shut down his practice as quickly as he could, consistent with doing things properly. He met with each of his clients and recommended other attorneys to represent them. Then he transferred their files to the attorneys they had selected. He cleared out his office and turned it over to his landlord.

What he did not do was tell his family. He thought he'd do that in person.

Pearl sold or gave away belongings that they would not be taking with them. She purchased riverboat and railroad tickets to Sapulpa, one of the larger towns in Indian Territory, and one with a station on a rail line.

They packed their bags and headed south to break the news and visit the Riggins family. The visit was brief but meaningful. The family warmed to Pearl.

"Why leave Des Moines? You worked hard to become an

attorney and soon established a good practice. Why walk away from it?"

"We were doing all right, but I had my sights set on doing better than all right. I think we'll do better in Indian Territory. It's a growing economy, unlike Des Moines. Attorneys who were going to make it there have already done so. There's not much chance of joining them.

Also, Indian Territory is exciting! It's an adventure. Des Moines passed through that stage a hundred years ago. It's staid now. I'm looking forward to spontaneity and chance."

"What are you going to do there?"

"Continue practicing law. I'll probably handle real estate transactions and do trial work, just as I was doing in Des Moines."

"But won't you miss your friends?"

"Of course. But there'll be people there. I suspect we'll make friends quickly, just as we did in Des Moines."

The questions and answers went on and on until the family was satisfied that Lew knew what he was doing and had made a wise choice, as he had done when he left home to go to college.

On the morning of the third day, the family gathered to wish them well.

"You're off to set the world on fire. I hope you do it, and I think you can. You're the most determined person I've ever known."

"Thank you, Father"

Lew and Pearl continued their journey to Indian Territory.

He felt determined and confident. She entrusted the future to her husband.

Thus it was that Lew left Iowa.

He never looked back.

5

SAPULPA

Sapulpa was the heart of Indian Territory.
Lew and Pearl decided to settle there. They purchased a house and rented a nearby office, an easy two-block walk between them.

The town was named after a Creek Indian from Alabama, James Sapulpa. He established a trading post in 1860. In 1886, rail service became available. By 1905, it had more than 4,000 residents.

There was a courthouse, a jail, a bank, a saloon, and a post office. There were schools, office buildings, retail businesses, farmers' markets, and residences. There were light industrial businesses—brick and other clay products, glass, cotton compressing, and storage and maintenance of railroad cars. It was an enterprising and growing community.

6

A SEPARATE WORLD

Sam was born in 1880 to Joe and Lena Mae Perryman. He belonged to a tribe that called itself "Muscogee," but which the white men called "Creek."

In 1834, pursuant to the Indian Removal Act, most Creek Indian tribes were ushered out of Alabama and Georgia and into Indian Territory. Not because they had done anything wrong, but because white people wanted their rich tribal lands in the Southeast. As compensation for the land they lost, the tribes received land in Oklahoma. What they received was not as good as the land they were forced to abandon.

Federal troops on horseback forced them to walk from the one place to the other, a distance of between 700 and 900 miles. Some of them did not survive the trip. Those who did never forgot it. They named it the "Long March."

They had been farmers. They had no formal education of the type that whites provided their children. They were ignorant of the white man's culture, traditions, business practices, and legal systems.

Sam's father had been part of the Long March. His mother

was born shortly after reaching Indian Territory. Their tribe settled on land in the western part of the Creek section of Indian Territory. Sam was born and grew up there.

Their community consisted of 200 acres. It was somewhat fertile land, mostly along a river. It was replete with game and wildlife.

The Tribal Town was in the middle of the 200 acres. People lived in small, one-room houses built of wood and clustered together. There was a larger structure, where people met, and a barn for draft horses, ponies for riding, and a few cattle. Smaller buildings served as storage and for farming implements.

Sam's father and other survivors of the Long March never stopped talking about it, especially to the younger generation. They wanted people to learn about it, to think about it, and to understand that the white man was no friend of the Creek. Whites took what they wanted and replaced it with something of lesser value.

The community was for the most part self-contained and isolated. There was very little contact with whites, who occasionally visited the community for purposes of trade. When the tribal elders deemed it necessary, men from the community went to town to sell produce and wild game and to buy farming equipment. From these encounters, the community learned what it knew of the white man's culture, traditions, and mores.

7
CHILDHOOD

Sam had an innate curiosity about the world and the things in it. He thought about how things worked. He tried to figure out why tools were built the way they were. He wondered why horses were made one way and cattle another. He watched cats jump and wondered how they did it. Almost anything tangible could trigger his curiosity.

He was a loner. He liked solitude. Except for sports, he preferred being by himself to being with other people. He liked to go into off into the distance alone.

His father worried about this. He thought It was important to be and stay in a united community.

"I've noticed that you avoid people. Why do you do that?"

"I guess I'm shy. And it never seems as if I have anything important to say."

"You might think about trying to talk more. The more you do, the less shy you'll feel. Give it a try."

Sam took his advice and talked more, but never got over feeling shy.

HISTORICALLY, the tribe had always had a communal economy. Everyone contributed to the welfare of the community.

The boys were required to perform chores.

Sam was assigned to farming when he was ten. By the time he was 12, he was plowing without supervision. He was ready to do it indefinitely. It had never occurred to him that he had an alternative. People in his community didn't complain.

One spring day he was plowing a large field, preparing it for planting. He had been assigned 20 rows. He put a stick in the ground to mark where he started. He finished the 20 rows, plus one extra.

Sam's father told him to always do more than he was assigned. That way he could always be sure that he was doing his part. Living communally, it wouldn't do to ask someone else to finish one's work.

Sam unhitched and tethered the horse for the next boy. Then he ran off toward the center of the town.

His father happened to see him running by. "Did you do your part, son?"

"Yes, Father. I did my part, and a little more, and now I'm going to join in a game with the other boys."

"You've not been at it for long."

"I worked quickly and steadily. I didn't stop to rest. And I did a good job. You can go see for yourself." And off he ran.

In addition to helping with farming, Sam, when he was a bit older, helped build and repair community structures. From the start, he was interested and enthusiastic about the work. He had a natural affinity for it. He learned quickly and developed considerable carpentry skills, doing work that pleased the elders.

Tribal elders were the community's governing body. They managed the community's assets and finances, adjudicated disputes, and established rules and protocols.

The community didn't provide formal education for its children. But the elders regularly talked to the boys, not the girls, grouped by age. They spoke of tribal traditions, the organization and governance of the tribe, and farming and hunting techniques.

In recognition of the inevitable assimilation into the white population, the elders decided to provide weekly English-language instruction to the children—both boys and girls. They developed techniques for the instruction.

An elder spoke English to the children, then asked them to repeat what he had said. He used objects, gestures, and motions to help the children understand and remember the words he used. He worked with one child at a time, while the others observed. This multiplied the impact of the lesson.

Elder: "I speak" (*hands palm to palm, flapping*) "a little" (*thumb and forefinger close together*) "English."

Child: "I talk small English."

It was a gradual process, built on repetition.

Elder: "'Speak'" (*hands palm to palm, flapping*), "not 'talk'."

Child: "Speak."

Elder: "'Little'" (*thumb and forefinger close together*), "not 'small'."

Child: "Little."

Elder: "I speak a little English."

Child: "I speak little English."

The elder overlooked the absence of the definite article. There was nothing similar in the Creek language. The children would have to learn articles by listening to whites. In the meantime, they would be able to get along, and the elder thought that was enough.

The children regarded the speaking lessons as play. Probably because of this, all of them learned to speak English, although with differences in fluency.

Sam learned to speak fairly well, better than some. He had developed his listening skills early, while observing wildlife.

Every other week, the elder taught reading and writing. In the absence of books, he used a primitive precursor to the chalkboard.

Reading and writing were much more difficult than speaking, but virtually all the children caught on, to some extent.

But not Sam. He tried hard, but had no aptitude. He could not see the patterns of letters that form words, so he could not distinguish words with similar appearances. And he could not see the patterns of words that form sentences.

For people like Sam who were having trouble with reading and writing, the elder designed a program to enable the child to write his or her name. He approached it as drawing. He would print the child's name in large letters, then ask him or her to copy it, as if it were a picture. By working at it, Sam got to where he could write something resembling his name. His letters were wiggly, but with a little imagination, a person could see them as "Sam."

8
SAM'S INTERESTS

Sam loved the outdoors. He was fascinated by animals and birds. He figured out the type of habitat that various species needed, where they lived, what food they ate, when they fed, when they bred, and when they gave birth.

He was most interested in deer and wild turkey. Because of the community's remoteness to dense populations, both species existed in profusion and could be observed year-round.

He often snuck out at night to listen to night sounds. Animals, birds, and insects contributed to what was sometimes a cacophony. He would be very still and distinguish the different sounds—coyotes, owls, frogs, crickets.

His favorites were whippoorwills, pileated woodpeckers, and caterwauling bobcats. These sounds had an otherworldly quality. Sam always got goosebumps when he heard them.

He also watched the sky. Again, because the community was isolated, he had a clear view of the moon and stars. He learned the phases of the moon and the positions of constellations in different seasons. One August night, he got so excited watching shooting stars that he ran home and woke his father. "Father, come quickly. The sky is falling."

His father thought, *Sam is really a special person.*

———

THE BOYS in the community engaged in a variety of sports: wrestling, tug-of-war, lacrosse, foot races. Sam liked all of them and was good at all of them. Next to foot races, he preferred the more physical of them—wrestling. He was big and strong for his age.

His father once asked about sports. "Which do you like the most?"

Sam didn't have to think about it. "Foot races."

"Why is that? Running seems more like work than fun."

"It feels good. When I hit my stride, I feel free, almost as if I were flying, as if the wind were carrying me. Sometimes I imagine that I'm a hawk, flying higher and higher until I just glide in the wind."

———

THE BOYS in the community hunted deer and fished. Sam was good at both. He liked to hunt and fish more than he liked playing with the other boys.

He fished throughout the year. Because of his curiosity and patient observation, he became a fine fisherman. He learned where fish held in the stream during different times of the year. He learned the best times to fish, influenced by the time of day and the season and the moon. And he learned what bait worked best in different locations and under different weather conditions.

Over time he taught himself to read the stream. Smooth water meant the bottom was flat and the stream was deep, good for larger fish. Choppy water meant that the bottom was rocky and the stream was shallow, good for younger, more gullible fish.

He hunted wild turkey in the spring. He did everything that he had been advised to do, but he saw very few turkeys and never got a good shot at one. He finally admitted it to a much older boy. "What you need," said the boy, "is a gobbler call. It'll attract gobblers as well as hens. On a good day you'll call up ten, 20, or even more. Then you can get a good shot."

"Do you have one?"

"I'll help you make one. Then I'll show you how to use it."

After that, Sam got more turkeys than the other boys.

In the fall, Sam hunted white-tailed deer. In August, he found the trails that the big bucks used, by following their scat. When the rut began, he put small piles of corn kernels near their trails. Then he awoke early and climbed a tree near the corn. When a buck came to the corn, he had a good shot at him.

Throughout the year, Sam hunted squirrels and rabbits. He was good with a bow.

Because of Sam's hunting and fishing skills, he could have harvested more game to sell. But he chose not to. His father had suggested a guideline, common in the Cherokee Tribe, that he internalized: "Don't take more than you need."

9

FIGHTING

The boys fought sometimes, even though the elders frowned on it. Sam never looked for trouble, but he was short-tempered and quick to take offense. He usually got the better of his opponent, unless the opponent was a couple of years older than he.

The fighting troubled his father. "Can't you control your temper? We all get angry, but we take a moment to cool off."

"I can't help it. I get so angry that I can't think. I start swinging."

"As long as you stay that way, it will bring you trouble."

His father was right. Fighting got Sam into trouble. His punishment was more farm work and other community chores. But he was too inexperienced to know how to control his temper. "You wanted to be alone," his father said. "Now everyone's leaving you alone. This isn't in your best interests."

One of the elders gathered all the adolescent boys and spoke pointedly about the concept of community and how it kept their tribe together.

"Some boys do more than their parts. For example, they do

more than the plowing assigned to them. This is good. It shows strong and responsible character development, and it shows caring for the community. It is much better for the boy and for the community than doing as little as needed to finish an assignment."

Afterwards, the boys talked about doing more work than assigned.

One ventured that doing extra made him feel good. "Doing more than required makes you feel like you belong."

Another said, "You get back what you give. If you do as little as possible, that's what you get back. But if you help others by doing more than assigned, you'll get help from others."

Sam said, "I do a little extra because I want to make sure I do my part. I don't want someone else doing my work."

Another said, "I think you do extra to show off. You want people to see how good you are. You're a blowhard and a showoff."

That did it. Sam was done talking. He responded with his fists.

He rushed the boy, surprising him, and landed a right to the jaw. But it glanced off, because the boy instinctively took a step to Sam's left. The boy was bigger than Sam and had longer arms. So, the boy led with a right hook aimed at Sam's face. But Sam was quicker and dodged the blow. That gave Sam an opening to land a series of body blows. The boy, to defend himself, tried hitting Sam's face but connected only to the sides of Sam's head, because Sam stayed close. Sam then landed a left uppercut, which stunned the boy and left him momentarily confused. Without hesitating, Sam threw a vicious right cross to the boy's face. The boy went down, nose bleeding. From the ground, he signaled that he had had enough.

Later that day, an elder sent for Sam to speak with him.

"You fought again."

"Yes, sir."

"There is no brawling here."

"He insulted me."

"How many times have your parents talked to you about fighting?"

"I don't know. Many."

"You must change. Do you agree?"

Sam was silent.

"Come. Answer my question."

"I don't think I can change. I'm made the way I am."

"I wish you would learn to see your anger and control it. It's one thing to fight at your age. But fighting when you're an adult won't be tolerated. Now's the time for you to learn to control yourself."

10

ALLOTMENTS

The Creek wanted to continue their centuries-old practice of communal land ownership. But the federal government wanted to avoid separate, communal economies in the midst of a capitalistic country. It wanted to impose individual land ownership. So, in a multi-step process that was complicated, controversial, and prolonged, the federal government distributed (or "allotted") tracts of what had been Creek tribal land to members of the tribe. The tribe resisted this, but lost once again to the will of the white man.

Uprooted, cast into an alien place, and deprived of their valued economic principle, the Creek learned early on what an American president would point out almost a century later—that there is no appeal from might.

AN ELDER CALLED a meeting for the men and for the boys of 16 and older.

"White men will no longer allow the tribe to keep its lands.

They require that all communities be disbanded. Their lands will be given to individual members of the tribe."

"How will it work? How will tribal members get the land?"

"Individuals who are of age must register as tribal members. That's the first step. Then each can state a preference for up to 160 acres that he would like to have. That's the second step. Then people in the land office decide who gets what."

"Where do we go to do these things?"

"The land office is in Muskogee."

SAM'S FATHER spoke to him later before the family retired for the night.

"You've been everywhere around here. Get on a pony," he told Sam. "Find land along the river, but land higher than the river. If it's higher, it won't flood. It should be good land to farm. Once you've found it, pay close attention to where it is in relation to bends in the river. When you get to the land office, ask to see a map of our area. Pick your land by the river bends on the map. Do you understand?"

"Yes, sir."

"The map will show squares, each consisting of 40 acres. You will be allowed to state a preference for four squares. Pick four that are all together so that they form a larger square."

"Yes, sir."

"Our community has horses, farm equipment, and some money. It will distribute these things to community members. In addition to some money, you should be able to get a pony, a draft horse, and maybe some equipment. You should have enough to start farming. Any questions?"

"No, sir."

"I suggest you start tomorrow. Getting to the land office before the others might give you an advantage."

Sam chose a pony to ride along the river to find the land that he would like to have. Then he rode to the land office. He registered and was able to identify the land he wanted from two bends in the river that showed on the map.

THE LAND OFFICE later notified Sam that he got 120 of the acres that he had selected. They consisted of three 40-acre tracts. In addition, he got a 40-acre tract by itself.

The 40-acre tract was some distance from the 120-acre tract, and away from the community's land, in an area that Sam didn't know. He tried to find the tract with a map, but the map confused him. With help from a neighbor, he finally found it.

He rode its entire periphery and four times through its middle, making a thorough assessment.

There is no tillable land. No topsoil. What dirt there is has a high clay content. And there are rocks. A lot of rocks.

The only trees are scrub oaks. They're stunted because the soil's so weak. They're twisted, wiry. Even though it's late spring, they haven't leafed out. They look sick.

It won't hold cattle, there's nothing for them to graze. There's no water.

There are no creatures.

Birds, animals, and hunters probably cross it occasionally, all searching for something to eat. But none of them stay. The land is too barren and rough.

This tract is no good for anything at all.

It's worthless.

11

COMMUNITY DISSOLUTION

Knowing that the community must dissolve, the elders had a duty to oversee the dissolution, including the division of assets. They met individually with all the male adults to learn about their tracts and plans. Some families received cash, some received farming equipment, and some got cattle or a horse.

As Sam was riding to perfect his right to an allotment, he thought about what was happening.

Life, as I know it, is ending. Everything I'm familiar with—other boys my age, sports, games, the local wildlife, the river—all will be gone. I'll have to start over.

I feel like I'm all alone. Without friends. Without the community where I grew up.

I wonder what the others are thinking. Live alone? Continue living with their parents?

I'm going to talk to some of them. If I talk to enough of them, maybe we'll come up with a plan, build each other's houses, work together.

If that doesn't work, what can I do about it on my own?

Maybe, just maybe, I can convince a group to create a small community on land that one of us owns. That would keep things close to what they have been. That would be great!

If I put my mind to it and talk it up, I bet it would happen.

12

MARRIAGE

But he never even got started. Unforeseeable circumstances intervened.

Sam's father introduced a subject about which Sam had not thought at all: "Son, have you considered getting married?"

Sam hesitated, caught by surprise.

No. Not really. Why do you ask?"

"I've been thinking about it."

"Yes? And?"

His father paused and gathered his thoughts. "Several things. You're a little young for it, but not that far off. And we now know that the community will have to disband. You know people. You know girls here. If you wait until you're older, you'll have to figure out how to meet girls. That could take a while. And, another thing. I 've been sick, and I would like to get to know your wife while I feel pretty good."

"Oh, father. You're not about to die."

"I know that. But I don't know how long I'll be able to get around and enjoy things."

"Okay. You've given me something to think about. I'll see where I come out."

"I appreciate that. And you might start by speaking to an Elder. They have a lot of wisdom about things like this."

A couple of days later, Sam sought out an Elder. After talking small talk, with several long pauses, Sam was able to address the subject. "Do you think I'm too young to marry?"

"You're on the young side. But I've seen people your age get married and it worked out well. The question is whether you want to get married. Do you have someone in mind?"

"No, not really. In fact, not at all."

"If that's the case, I'm surprised you're thinking about it. What gave you the idea?'

"My father suggested it. He thinks that I might want to find someone in our community, before it disbands."

"That's interesting. And not a bad idea at all. You're basically a loner. You don't relate easily to others. A young woman in our community would know about you, even if you had not had much to do with her. Outside our community you might not find a wife."

"I think that's what my father is concerned about."

"And there's another consideration. You now have land. If you get married before you leave the community, your wife will be involved in decisions that affect both of you. Things like where you live and what kind of abode you'll have."

"That's a good point. I hadn't thought of that."

"Let me think about it. Come back tomorrow, and I'll see if I can suggest a starting point for you."

Come the next day, Sam returned to the Elder.

"Sam. I'm glad you came. I wasn't sure that you'd follow through. It's a touchy subject."

"You're right. I had trouble sleeping last night. I was debating whether to drop the whole idea."

"I'm glad you didn't. I've thought about the young women in the community, and I think I've come up with a good suggestion. Do you know Lucy Harjo?"

"No. But I know who she is. I think she's very pretty."

"She's your age, maybe a little younger. I doubt that she's spoken for. And she's a good person with a nice disposition and with a lot of common sense. On top of that, she's worked for a couple of years with the Elders who keep track of the community's finances and who make decisions on legal matters. She has a lot of experience reading English and is good with numbers."

Sam fidgeted.

I don't know what to say.

"Sam. Are you all right? You look confused."

"Yes, I am confused. I don't know what to do next."

"Then let me help you. You've got to figure out ways to get to know her. Perhaps your parents can suggest things to do together. But one way or another, you've got to get to know one another. That's the only way to see how you feel about each other. And if you find that you don't feel something for her, it would be a mistake to go further."

Not long after, Sam's father and Lucy's father arranged for the Harjos to have a meal at Sam's house. The purpose of the meeting was obvious to everyone.

Then Sam did his part. He arranged walks in the woods with Lucy. He even took her fishing and hunting. And he brought her to sports events, in most of which he participated.

When Sam had made up his mind, he discussed marriage with Lucy. They both wanted it. But she said he would have to get her father's permission.

Requesting permission makes me nervous.

Once inside the Harjo's home, he was awkward, in addition to being nervous. "I came to. I mean. I came to. Let me explain."

Mr. Harjo was understanding.

"I know why you're here, Sam. And I understand why you're nervous. I, also, was nervous when I asked to marry. And I know that you, particularly, find it difficult to talk about sensitive subjects. But you can relax. I'm very pleased to have you marry Lucy."

Mr. Harjo could see that Sam was relieved. "But I want you to promise me something, Sam. While you're a fighter, and a little rough around the edges, I've never heard anyone say that you're unkind. And you come from a good family, so I'm sure you'll know how to be a good husband. Can you promise me, Sam, that you'll take care of Lucy and be kind to her? She's a wonderful person, and she deserves a happy marriage."

"Yes, sir. I gladly promise that. You have my word on it."

THEY HAD a traditional Creek wedding ceremony attended by everyone still living in the community, plus some who came for the event. Afterwards they went to an abode arranged by Lucy's father. It had a bed in it. And both parents had lent other necessities for the new couple to take up residence.

That night, Sam and Lucy lay together in their marital bed. It was an entirely new experience for Sam. He had not known physical intimacy with a woman. He was awkward, but Lucy was patient. With her understanding, and with a few-unspoken directions, things went well enough, although not smoothly.

13

NEW HOME

I want to build a house on the 120 acres. I want it to be on the edge of the land closest to the road and farthest from the river. I don't want to ride through cropland to get to and from my house.

He and other young men from the community formed a group to help each other build houses on their newly acquired land. They agreed on a simple standard design of two small bedrooms, a kitchen and a common area. There would be a peaked roof with a 30-degree pitch on each side.

The group agreed to put up all outer walls, including standard windows and doors, the roof, flooring, framing for ceilings and for interior walls between rooms, and an outhouse. The owner would finish out the ceilings and inner walls and take care of exterior and interior painting.

To hold down costs, the group acted as a unit to negotiate best prices for lumber, nails, and shingles. The group used part of the money that the community had given them to pay for supplies.

The houses were solidly built to last a long time. The group had been careful to buy lumber that, if properly cared for, would last virtually forever.

After the group had fleshed out each other's houses, Sam first painted his house's exterior. Then he whitewashed the interior walls and ceilings.

At this point, I have a serviceable house. It's simple but simple is good. It'll be livable. I'll add the finishing touches over time.

He and Lucy moved in. She quickly demonstrated talents for making the house feel welcoming and comfortable.

Realizing that they would need heat in the approaching winter, Sam searched around the then-disbanded community and found a pot-bellied stove in good condition that someone had left behind.

"I'm sorry the house isn't done. I'd hoped that it would be further along when we moved in."

"We had to move from the community, and you've provided a house. It's our home. We'll enjoy it as we finish it. I look forward to helping with that."

AT THE INVITATION of Sam and Lucy, his mother and father also moved in. They had to find a home because they too were displaced.

Lucy made the best of hosting her in-laws. She with Sam's mother beforehand. Sam's mother wanted to make sure Lucy thought they were pulling their weight.

"You can help me in the kitchen," said Lucy.

"I'll look forward to that," said Sam's mother.

"And you can keep an eye on Sam's father. He needs attention every day. We'll all pitch in, when it's needed. But it would be a relief to Sam and me to know that you're watching him."

"I wouldn't think of having others do it for me. It's my job as his wife."

"Good. And one other thing while we're talking. I haven't mentioned this to Sam, and I'd appreciate your keeping it between us. He does not understand money. That's a job that I'm going to have to handle. We won't be able to plant for a month or so. The harvest will be several months off. I don't know that he and I have enough money to carry us until harvest. So if you could think about pooling your money with ours, it would relieve my concerns. Maybe you could talk to Sam's father about it."

"There's no need for that. I can speak for both of us. I think it's a great idea. It would make me feel better about moving in with you. We would be carrying our own weight."

Lucy was so happy that she hugged her mother-in-law. "Thank you. Thank you. I'm so happy that we can be honest with each other."

AFTER HIS PARENTS MOVED IN, Sam turned his attention to building a simple barn with three walls, the open side facing south.

And as everyone settled in, Sam developed daily chores. He hauled water from the river and chopped wood for cooking and heating.

Lucy took him aside. "You're working very hard. I appreciate it, and I'm proud of you. You've changed."

"It's very different from growing up. Then, for the most part, I could do whatever I wanted; there were others to take up the slack. But now, I feel responsible for three people."

14

SAM'S FATHER

Sam's father slept most of the time. He died one night. It was a quiet passing.

Sam's mother and Lucy discussed where to bury his remains. The discussion was a formality, because Sam's mother, the most senior family member, had the final say.

"Do you think we should start a family burial ground here?" Lucy asked.

"I think it would be more fitting to use the community's cemetery. After all, he spent virtually his whole life there."

"What will happen to the cemetery after the community's gone?"

"The elders say that the government has promised to leave it alone. That was part of the agreement."

"Then we should use the community cemetery."

THEY HELD a traditional Creek funeral at the community cemetery. They were pleased that people came from their new locations to participate.

I'm impressed with how much people respected my father. While growing up, I had been so self-absorbed that I had not observed the reasons for their respect. The more I think about it, the more I wish I had known my father better.

I really should talk to Lucy about this. But I know I can't bring myself to do it. I'm not sure why, but the idea of the conversation frightens me.

15

FARMING

Sam settled into his new life as a farmer and a husband. He worked hard, doing all the work by himself.

He had beginner's luck with the weather. Not too hot. Plenty of rain in the spring and summer. And dry weather for harvesting. So, Sam had a good harvest. He sold his crops to several people and gave the cash to Lucy after each sale.

I like what I'm doing and working for myself makes it all the better.

As he had time, he fished and hunted. This provided additional table fare for the family.

In the winter, he worked on the house and kept up his equipment.

Lucy sometimes worried that he was pushing himself too hard.

Sam said nothing about how he felt concerning his new roles.

I'd like more time to hunt and fish. But this is all new to me, and I want to make sure I do my part. Certainly, Lucy is doing her part. To be fair, I need to do mine.

THE FEED STORE in Sapulpa gave Sam credit for seed. After the harvest in his second year of farming, he went to Sapulpa to settle his account.

As he was leaving the store, the clerk said, "You've had a good year; why not celebrate?"

He thought, *White men ask questions they don't expect answers to.*

"Treat yourself to a drink of whiskey."

I've never had whiskey. Wonder if I would like it.

"Where would I get whiskey?"

"Down the street. A place called 'The Lazy Horse.' It has a sign out front with a horse painted on it."

16
LEW AND SAM—FIRST MEETING

"Good morning, sheriff."
"Mornin', Lew. Thanks for coming over."
"What's going on?"
"Another drunk Indian in the lockup."
"Is he drunk now?"
"He was yesterday afternoon, when I arrested him."
"How can I help you?"
"When I got here this morning, I told him he faced some serious charges and could even go to jail. I told him he had a right to talk to a lawyer. He acted like he didn't want a lawyer. I explained how a lawyer might help him and asked if he wanted one. He thought about it and nodded. Since you're the only lawyer in town who'll represent an Indian, I sent Billy over to fetch you."
"I see. I'm happy to help him, if I can. What'd he do?"
"He pitched a ruckus over to the Lazy Horse. Jim wouldn't give him a third drink, so he started arguing. When Jim got mad and told him to get out, he flew off the handle. He started yelling and throwing things around. Then other customers tried to force

him out the door, and he got rough. He's big and strong. And he's tough and knows how to fight. He cold-cocked two of them. He woulda done more damage, except that Jim got him from behind with a pickaxe handle to the head. I bet he's got a headache this morning."

"Did he see the sign?"

"Don't know. When I asked about it, he just stared at me. Wouldn't say a thing."

"Maybe it offended him. Were the other customers white?"

"All of them."

"Did you arrest any of them?"

"Now, Lew," said the sheriff, looking him in the eye, "you know better than to ask that."

Pause.

"When's the arraignment?"

"Tomorrow. Judge Mayfield's in town then."

"Does this Indian have any money?"

"Probably not. Like the rest, he probably only has some worthless land around here. Can't grow anything on it. Only good for grazing. Need a bunch more of it than any of them has, to make any money."

"Can I talk to him?"

"You know the way, but I have to go with you. Rules, you know. But if he wants to talk to you, I'll leave the two of you alone."

As they walked down the hall, Lew pointed to a calendar on the wall. "You ought to get a 1905 calendar, so you can know for sure what day of the week it is."

"Doesn't seem worth the trouble. One day here is about like the next."

When they reached the cell, the sheriff unlocked the door and he and Lew went in. The sheriff introduced them to each other.

"Sam Perryman, this here's Mister Lewis Riggins. You wanna talk to him?"

Sam nodded.

Lew thought, *He seems subdued, but I need to question him.*

"You ever been in trouble with the law before?"

Sam slowly shook his head.

"Tell me what happened last night."

Sam spoke slowly and haltingly, giving an account that was about the same as the sheriff's.

"Did you see the sign about Indians?"

Sam pursed his lips and stared at the floor, saying nothing.

Lew paused for a minute or two while he was thinking.

I'm going to let it drop for now.

"I think I can help you. But we need to talk about my fee."

I learned while reading law and from my law practice in Iowa a basic rule for taking on an engagement to represent a criminal defendant. Always collect your fee before beginning the representation. Otherwise, you probably won't get paid for your work. If the defendant is convicted, he'll blame you and refuse to pay. If he's acquitted, he'll convince himself that he didn't need your help and refuse to pay.

"Do you have any money?"

"I sell deer meat."

"How about land? Do you still have your allotment?"

"I have a 40-acre tract and a 120-acre tract. I farm the 120, and I live there. I've only seen the 40 once. Not good for anything. Soil's too poor to grow anything, so it's no good for crops, or even for cattle."

"Where's the 40?"

"Almost due west of here. Not sure how far."

I know nothing about the tract. I'm not even sure where it is. But I have to make a decision. Either accept the land for what it is or do the work without an up-front fee, which would likely mean no fee.

"Would you be willing to transfer the tract to me, for my fee?"

"Yeah. It's no good to me. When can we do it?"

"Later today. The arraignment, when a judge reads the charges against you, will be tomorrow."

Pearl, who was doing Lew's secretarial and clerical work, went to the title examiner's office, where she found the legal description of the land. Then she prepared a deed from Samuel Perryman to Lewis B. Riggins.

In the sheriff's presence, Lew explained to Sam the effect of the deed and the reason for it. Sam acknowledged that he understood. So Lew signed the deed and Sam drew his signature the way the elder taught him to do. The sheriff was the witness.

PROMPTLY AT NINE o'clock the next morning, Judge Mayfield entered the court room. The bailiff called the docket, which began with Sam's arraignment.

The judge read the charges. "Mr. Perryman, you are charged with aggravated assault, drunk and disorderly conduct, criminal trespass, and malicious destruction of property. These are serious charges that, if you are convicted of one or more of them, could result in fines or imprisonment or both. Has your attorney explained these charges to you?"

"Yes, sir."

"Do you understand the charges?"

"Yes, sir."

"How do you plead to the charges?"

"Not guilty."

Looking at the prosecutor, the judge said, "I need to decide about bail. What do you have to say?"

"Your Honor, I would be satisfied with releasing Mr. Perryman on his own recognizance. That's the usual arrangement for Lew's Indian clients, and he always sees that they come to trial."

"I'll enter that order. I'm going to set the case for trial three weeks from today."

The judge told Sam he could go home. So, Sam promptly got his horse and started riding.

Later, Lew took Sam's deed to the Interior Department office, which approved it. Then he filed it with the recorder of deeds, for the public record.

17

SAM AND LUCY

Sam arrived home early in the afternoon. Lucy ran from the house to meet him. *She's agitated, which is unusual.*

"Fancy seeing you here. You were supposed to be home Tuesday evening. Here you come, ambling in, Thursday afternoon. Whatever happened to you? I've been worried sick."

"It's hard to explain."

"Come on, Sam. Talk. I'm your wife and deserve to know."

"I got done with my business. Then I went into a nearby saloon. I had two whiskies and then asked for another."

"Oh, no."

"But the saloonkeeper acted like he hadn't heard me. I got angry and kept asking for another whiskey, louder and louder, and he started yelling at me. Before I knew it, two men rushed over and one tried to hit me in the face, but I ducked it. Then I came up fighting and knocked both of them out. Someone hit me from behind with a club or a board. When I came to, I was in a jail cell."

"Oh, dear. Fighting again. Were you drunk?"

"No. They serve whiskey in little glasses," said Sam, holding his thumb and forefinger about two inches apart, "about this big."

"Why did you go out of your way to drink whiskey?"

"I had heard people talking about whiskey, but had never had any. I wanted to see what it was like."

"Why did those two men start a fight with you? Did you say something to them that made them mad?"

"No. I didn't say anything to them."

"The whole thing is strange. Are you telling me everything?"

"Yes."

"What happened when you came to in the jail?"

"Nothing. I was alone. It was night. My head hurt, and I didn't feel good. So, I just went to sleep."

"And then?"

"A man woke me in the morning, bringing me something to eat. I was hungry, so I took it. After I finished eating, the man came back with another man. The first man pointed at the second man and said his name was Riggins. Then the first man left."

"Who were these men?"

"I don't know. But Mr. Riggins seemed to know what was happening. He asked me some questions but I didn't answer them because I was afraid. Then he told me that I was in trouble, and he could help me."

"Help you with what?"

"I'm not sure, but he said I had to see a judge. Then he said he charged for helping in a situation like this. I told him I didn't have much money. Then he asked if I still had any of the land I was given. I said yes, and he asked if I would give some of it to him for helping me. I thought for a minute and said I would give him 40 of my acres. But I also told him that it wasn't very good land."

"You agreed to give away some of your land just to get this

man to help you with something that you didn't understand? Do you know what you just did?"

"I was afraid. I didn't understand what was happening and what people were saying. But I did understand that I was in trouble and that I might have to be in jail for weeks or months."

"What happened next?"

"Mr. Riggins came back with a paper and with the other man. Mr. Riggins told me to write my name on the paper. So, I drew 'Sam' and then the other man wrote something.

"The next morning, Mr. Riggins and I walked to a big building across the street. He seemed to know where he was going, so I went with him. We ended up in a big room with several men in it, including one in a black robe sitting behind a long table. Mr. Riggins told me that was the judge. Then Mr. Riggins and another man, who Mr. Riggins said was the 'prosecutor,' went to the table and talked to the judge. After a minute or two the judge looked at me and told me that I could go home, but he said I had to come back in three weeks for my trial."

"You must have done something wrong in the saloon. It was probably the fistfight, even though you thought you were defending yourself. You've been fighting all your life. I thought you'd gotten over it, but apparently you haven't."

"If someone tries to hit me, or even insults me, I start fighting without thinking about it."

"I'm deeply disappointed. You fool. Your mother will be furious. Go clean yourself up before she sees you like this. I hope you've thought about what you're going to tell her. She was also upset and worried when you didn't come home Tuesday."

A hangdog look came over Sam. "I can't talk to her about it. Would you mind explaining it to her?"

"Okay, I'll do it. But I want you to remember that she's part of our household and things affect her just as they do me."

"Thank you."

"Think about your fighting. You've got to learn to control your anger."

Sam swallowed hard. "I agree. I need to control my anger."

FOR THE NEXT THREE WEEKS, Sam did everything he could to stay in Lucy's good graces. But a pall hung over them. They didn't understand the judicial process, so they weren't able to evaluate the likely outcome.

On the appointed day, Sam left early in the morning for Sapulpa, hiding his panic and fear.

18

THE TRIAL

After the jury was empaneled and sworn, the judge explained the charges against Sam and the consequences of conviction.

That's scary. I've heard these before, at the earlier meeting with the judge. But now, at my trial, I feel panic.

Who will take care of the house, for Lucy and my mother? Who will prepare the equipment for next year's planting?

Overcome with fear about the future. Sam was having trouble paying attention to anything going on in the court room.

He felt all eyes were on him, and some men in the jury looked at him as if they were disgusted or disapproved.

The prosecutor called the saloonkeeper as a witness and asked him to describe what had happened. After he finished, Lew questioned him.

"Had the defendant ever been in your bar before the afternoon of the fighting?"

"No, sir. That was his first visit."

"When the defendant asked for a third drink, did you explain the rule to him?"

"No, sir. I can't say that I did."

"Jimmy, would it be fair to say that you have a quick temper?"

"Well, yes. I guess everyone here knows that."

"And did you lose your temper with the defendant?"

"Yes, I did."

"Why?"

"Because he was arguing with me about a third drink, when the rule is only two."

"For the jury, would you would you explain the rule?"

"Yes. It's simple. Indians can only have two drinks."

"Is the rule displayed in the bar room?"

"Yes, it's on a sign, on the wall behind the bar."

"You got angry because the defendant was acting as if the sign was not there?"

"Yes. I couldn't see why he was arguing, and the arguing was bothering other customers."

"Do you remember whether you swore at the defendant?"

"Yes. I do remember, and I did swear at him."

"Did you, in fact, use the words, 'fucking Indian'?"

"I did. I'm not proud of it, but I did."

"Is that when the defendant started shouting and throwing things?"

"Right after."

"No further questions, Your Honor."

The prosecutor said, "Your Honor, I'm finished presenting my case."

Lew began to present his case.

"I call Sam Perryman to testify."

Hearing his name brought Sam out of his stupor.

Seeing that Sam was not moving, Lew motioned to him to sit in the witness stand. The bailiff approached Sam with a Bible.

Lew put Sam's right hand on the Bible and raised Sam's left hand.

After Sam was sworn, Lew addressed the judge. "I wish to make an autoptic proference."

The prosecutor jumped to his feet. "Your Honor, may we have a conference at the bench?"

"Yes. Please approach."

The conference was conducted with lowered voices.

"Lew. What in the name of hell is an autoptic proference? Is this some kind of trick?"

"Not at all. It's a well-recognized type of evidence. See Chapter 39 in *Wigmore on Evidence*. It's an exhibit that speaks for itself."

The judge looked puzzled. "Let's have a look at it."

Lew picked up a bundle from the counsel's table and returned to the bench, where he unwrapped it, revealing a sign.

The prosecutor would have none of it. "Your Honor, I object. It's a trick. It has no probative value. And what it says isn't necessarily true. It could really confuse the jury."

The judge thought for a minute. "I agree with you on both counts. But I think I see what Lew is doing. I'm going to take a chance and allow it into evidence. It could be reversible error if I don't."

People then commenced to speak in normal voices.

Lew took the sign to the court reporter. "I'd like to introduce this into evidence. Please mark it as Defendant's Exhibit One."

Lew then walked to the jury box and held up the sign for the jury to see. It read, "TELL YOUR SIDE OF THE STORY, AND YOU WON'T GO TO JAIL."

After the jurors had time to read it, he carried the sign to the witness stand.

"Sam, would you please read this sign for the jury?"

There was a long silence while Sam stared at the sign.

"Sam, I'll repeat the question. Will you please read the sign for the jury?"

Sam fidgeted in the chair. "No."

"Why not, Sam?"

Silence again, with Sam staring at the floor.

"This is important," Lew said. Then, moving closer to Sam, in a lowered voice he said, "This is no time to hide something. Just tell the truth."

After a moment, Sam said, speaking softly, "I can't read it, because I can't read." Then, raising his voice, he quickly added, "It's not my fault. I'm not dumb. The tribe taught us English. I learned how to speak it pretty well. But no matter how hard I tried, I couldn't do the writing and reading.

"The elder who taught us said there was something wrong with me that was keeping me from it. He said I was born with it."

There was silence in the courtroom. No one moved. Everyone seemed embarrassed to have learned Sam's painful secret.

The judge broke the silence by calling Lew and the prosecutor to the bench. Looking at the prosecutor, he said, "I ask you to agree to a directed acquittal. The sign in the Lazy Horse is critical to your case. Jim got angry because he assumed that the defendant had read it and was arguing with it. And Jim's anger and his resulting profanity provoked the defendant. Knowing that the defendant can't read, the things he did were a justifiable response. I can't let the jury convict him of any of the charges."

The prosecutor thought for a minute and then agreed with the judge.

The judge turned in his chair and addressed Sam. "I'm acquitting you of all charges. This means you are free to go. You will not be punished. But I feel constrained to advise you to stay away from alcohol forever. It's not your friend. It got you into trouble this time, and might again. Think about that before you ever take another drink."

"I can go now, right?" said Sam.

The judge said "Yes". He then pounded his gavel and announced, "Adjourned."

Before Sam left, he thanked Lew.

What a relief. I'm a free man. I can go on with my life as it has been.

As the courtroom cleared, the prosecutor approached Lew and said, "Clever, Lew. Very clever. And now Indians will overrun this town."

19

SAM—ADJUSTMENTS

There's Sam, Lucy thought. *I'm surprised and relieved to see him return this soon.*

They spoke out of her mother's hearing. "Tell me what happened, Sam."

"It didn't last very long. Mr. Riggins and the prosecutor were both there. And there were other people in the room. Twelve were seated together over to the side. When the judge came in, everyone stood up. He is called 'Honor.' People kept looking at me and pointing at me. I was confused and afraid."

"Can you tell me what happened?"

"Twelve people went to a box, one by one, and sat down and then answered questions from the prosecutor and Mr. Riggins. Then Mr. Riggins and the prosecutor had a talk with the judge. Mr. Riggins showed them a board with words on it. The prosecutor seemed upset. Then Mr. Riggins showed the board to the people sitting in the box. Then he showed it to me. He asked me to read it. He kept holding the board and looking at me. After a moment, he asked me why I didn't read the words. I didn't want to answer, because I didn't want the people in the room to know, but he told me I had to. So, I told him I couldn't read. Then the judge looked at me, then at Mr. Riggins, then at the prosecutor. Then the three of them talked. And afterwards, the judge looked at me and said I could go home."

"That was it? You had to go all that distance just for that?"

"I guess so. I still don't understand it, but I guess so."

"It doesn't make sense. But whatever happened, one thing is clear. You made a mistake. Fighting is always a mistake. You've got to learn to control your temper."

"I got a strong feeling that the people in that room didn't like me. I don't know why. But there it is."

After a hesitation, Lucy said, "It's probably because you're an Indian. And that could explain why the two men tried to start a fight."

That puzzles me. Lucy is full of surprises concerning this whole thing.

"I don't understand. If the white man doesn't want Indians to be around him, why did he move us here from Alabama?"

"I don't know, but I suspect that white people would rather that we not be here. And I suspect that that was the cause of the whole incident."

After a silence, Lucy spoke again. "I hope you learned something. After all, you lost 40 acres over it."

"Lucy, I lost nothing. That land is worthless. There's hardly any topsoil. It's almost all rocks, trees, and clay. It has no water. We couldn't plant on it. We couldn't graze on it. The loser is Mr. Riggins. He gained nothing from his efforts. I told him that the land had no value, so I don't feel bad for him. I can take you to the 40 acres and let you see for yourself. It would take most of a day, but we can do it if you want."

"No, thanks. I'll take your word for what the land looks like. But land is permanent. After all is said and done, it remains; it never goes away. You're always better off having it. There's no tellin' how it might have benefited us, if you had held on to it." Lucy was getting irritated. Her voice was rising. "Are you listening to me? You made a mistake, giving your land to Mr. Riggins."

I don't like Lucy's angry look and the tone of her voice. I don't know what to say.

He said nothing.

After thinking about it, and after remembering the judge's admonition, Sam reached a conclusion.

The problem was not fighting. The problem was whiskey. Without the whiskey, there would have been no fighting.

I'm not going to tell this to Lucy. She would disagree, and I don't see any point in arguing about it.

He never took another drink of alcohol.

IT TOOK Sam and Lucy two or three weeks to get over the incident. They talked about it a lot. Eventually, they tired of it.

They never again spoke of the 40 acres.

20

STATEHOOD

The Sapulpa acting commissioner called a morning meeting at the Lazy Horse.

"We're all here. I'm sure everyone knows Lew. So, Lew, explain why you suggested this meeting."

"I know some of you believe differently, but Indian Territory and Oklahoma Territory would be better off as a single state. Local governance would give us more control over our lives and businesses. We'd elect our own governor and have more influence in Washington.

"There's a process for becoming a state. The starting point is for us to petition the President to ask Congress to adopt appropriate legislation."

After a discussion, the group decided to proceed. Lew volunteered to send the petition to the President, signed by members of the group.

IN JUNE 1906, Congress passed an act that permitted the creation of a new state fusing Oklahoma Territory with Indian Territory.

"You're a little late this evening," Pearl said to Lew.

"I stopped by the courthouse to register my candidacy to be a delegate to the constitutional convention."

"I thought you were interested in law, not politics."

"I'm not interested in performing a public service. As a delegate I will get information, so I can know what's happening."

"What difference will it make?"

The constitution will designate the borders of the state's counties. And each county will **vote** on a county seat. I want to know whether Sapulpa will be a county seat. That's a vital concern for me. My law practice depends on it. If Bristow and Sapulpa are both in the same county, I stand to lose a lot if Bristow is selected as the county seat."

"The court will be farther away, but what else?"

"The court's the important thing. I go to the courthouse almost daily, and sometimes more than once a day. Also, the county records are here. If I had to go to Bristow every time I wanted to check on a land title, I would lose a lot of time. And I couldn't charge a client for the time."

"I see what you're saying. The court's present location is a huge efficiency for you."

THE CONVENTION, with 55 delegates, including Lew, was held in Guthrie, beginning on November 20, 1906. It lasted, off and on, for more than nine months. As a final action, the convention set September 17, 1907, as the date that the public would vote on whether to approve the final draft of the state constitution.

Seventy-one percent of the voters approved.

President Theodore Roosevelt declared Oklahoma a state on November 16, 1907.

THE CONSTITUTION DEFINED the county including Sapulpa and named it "Creek County." To Lew's chagrin, it also included Bristow.

"What can you do about it?"

"I've got to temporarily lighten up my practice and spend some time in Bristow."

"What will you do there?"

"Talk to people. Point out the advantages of selecting Sapulpa as the county seat."

"And what are those?"

"People are used to the court and the records being in Sapulpa. Change that, and people in Bristow have to adjust. And Sapulpa is closer to Tulsa than Bristow. Also, Sapulpa has a train depot. All of that adds up to more growth for the county than would happen if Bristow were the county seat."

Lew spent the day of the special election in Bristow, at its polling station.

Sapulpa was selected.

PEARL HAD BEEN CONCERNED about Lew's political efforts. He'd now made some political enemies.

"People will soon forget. And I got what I wanted. With statehood, and with Sapulpa as the county seat, things have stabilized. Our continued success is assured.

21

LEW'S OIL BUSINESS

When word of the Perryman trial got around, Lew's practice grew. People in trouble wanted him to represent them. And he continued to be the only attorney in Sapulpa who would represent Indians.

Indians typically had no cash to speak of, so many of them deeded land to Lew as their fees, as Sam had done.

Pearl assumed the role of secretary, as she had in Iowa. As he got busier, she took on the duties of an office manager. She had good business sense.

Lew had become comfortable about their move to what was now Oklahoma. He was confident that he would prosper in his law practice. And he had gotten to know most of the men in town. This helped his practice and also opened opportunities for a social life for Pearl and him.

Over time, Lew and Pearl became respected and liked in social circles. The social structure was not nearly as developed as what they had been used to in Des Moines, but they participated in what there was.

As Lew learned more about the town, he developed a practice of dropping by the Lazy Horse for a drink on his way home. One evening, he joined a table of prominent businessmen.

"Hey, Lew. How are you doin'?"

"I'm doing fine, thank you. How about you gentlemen?"

"We're talking about people who are drilling for oil west of town, near Drumright and Cushing. Some of them struck oil."

"How'd they go about that?"

"There are people in that business. You could talk to a couple of them."

In his usual fashion, Lew absorbed the information about drilling for oil. He became interested enough to look through the deeds earned through his services. Perryman's deed covered land very near Drumright.

That night he broached the subject with Pearl. "I've been talking to people about oil. A couple of guys struck oil west of here, about 30 or 40 miles. Kind of interesting, isn't it?"

"I know you enough to know that if you're talking about it, you'd like to give it a try."

"I've looked through my Indian deeds. The one from Perryman covers land that's promising. It's near some successful wells."

"I suppose this will cost us money."

"I agree that I need to find out what things cost."

"That would be a good starting point. We don't have money to throw away."

"But a well could pay off."

"Maybe it won't. I've read about people ruined for life after they found only rock."

"There's some drilling going on about a day's ride away. I'd like to see it first hand."

———

A FEW DAYS LATER, Lew arranged for a horse, tack, and saddlebags. He woke early and started riding. In the middle of the afternoon, he could see a drilling rig. His excitement grew as he got closer.

I can feel something in my bones. I'm onto something big.

"GOOD AFTERNOON. I wonder if you could take a moment to answer some questions?"

"Who are you?"

"My name is Riggins. Attorney Lew Riggins. I'd like to learn how drilling works. I might be interested in doing some."

"I can help with that."

"This is a cable-tool rig. I'm in charge of it. They call me a 'toolpusher'.

"It's really pretty simple. Come over this way." Pointing at the ground, the man said, "You need a bore to get to where the oil is. That's what drilling creates."

He led Lew to a 30-foot cylinder of solid steel, about eight inches in diameter. "This bit has sharp teeth on the end. We lift it up, above the bore, and then let it drop. On its way down, the teeth clear away dirt and clay and they break rock into small pieces."

Lew was thinking,

It's just gravity. The drilling is accomplished with gravity! It's that simple. The only function of the derrick and the other equipment is to raise the bit above the hole.

Lew thanked the toolpusher and got back on his horse.

LEW WENT to his hotel to rest and think. While there, he struck up a conversation with a well owner.

"I'm wondering how you get started on a well."

"You find a geologist who works in the area of your land. He'll first reach an opinion on whether drilling is likely to be successful. Then he'll estimate the costs of drilling."

"How do I find a geologist?"

"Dan Harrington has an office in Sapulpa. He's not always there because he spends a lot of time in the field. But if you leave word that you want to talk to him, he'll find you."

"Does he have a good reputation?"

"If he works in the area of your land, you want to use him. You won't find any better."

LEW WAS TOO excited to sleep that night. He finally gave up at dawn. He got something to eat and started his journey back to Sapulpa.

Then he told Pearl all about visiting the drill site and how it worked and about locating a geologist to get a cost projection.

"I've looked through my Indian deeds. The one from that Perryman guy covers land that's promising. It's near some successful wells. So, I'd like to look into it a bit. What do you think?"

"That wouldn't hurt anything. It'd be good to learn more about the costs involved. We can probably afford to risk some of our savings. But there's something new that we need to take into account."

That sounds ominous. I hope everything's OK.

"What's that, Pearl?" Is everything all right?"

"Oh yes, nothing to worry about. But I have some news. I saw Doctor Riley today. I'm pregnant."

"That's great. I'm excited. We've been trying for a while. And I've been wondering if it was going to happen. But I haven't wanted to bring it up, for fear of frightening you."

"Well, to be honest, I've been wondering the same thing."

"But what does pregnancy have to do with oil?"

"It's just that we have to watch our spending. There'll be some extra expenses. One for sure—we'll have to hire temporary help to replace me in the office."

"Yes, of course. I see what you mean."

THE NEXT MORNING, Lew found Mr. Harrington's office. By luck, he was there.

"Mr. Harrington, may I interrupt you?"

"Certainly. Maybe you can introduce yourself."

"I'm Lew Riggins."

"May I call you Lew?"

"Everyone does."

"Call me Dan. How can I help you?"

Lew explained his profession, where his office was, the 40 acres Sam Perryman signed over to him as a legal fee, and what he had learned the day before.

"What I'm looking for now is a cost estimate."

Dan leaned back in his chair. "Sensible. Too many people get into drilling without thinking.

I charge by the hour to determine whether you have a good prospect. I have to do that because each property is unique. Some require a lot of work; some don't. But I can give you a pretty good estimate after I see your deed. But remember that this is all a risk. If you don't have a good well, you won't recover my fees or the drilling costs."

About a week later, Harrington went to Lew's office and gave him the estimate. That night, he and Pearl discussed it. They decided to proceed, but she cautioned him, "Lew, this is tight, but we can afford to try it once. Let's hope it works."

The next morning, Lew told Harrington to proceed.

Weeks later, a boy from the telegraph office delivered an envelope to Lew. Lew thanked him and opened the envelope. Then he left his office and went home.

How I should say this?

He opened the door and called out. "Pearl?

We're rich!

It's a gusher."

After Lew learned that he had a good well, he went to Harrington's office to discuss it.

"When can I start selling the oil?"

"There are a number of steps that need to be taken before you can sell any oil. Do you want me to take care of these?"

"Yes, definitely."

"But I think you should personally hire the pumper. You will be his boss, so its like a job interview."

"What's a 'pumper'?"

"That's the term we use for the person who checks the lease every day. If something is wrong, he sees to it that the problem is resolved. It's the pumper's job to make sure that everything stays in working order all the time."

Lew talked to owners of nearby wells. Based on their information, he hired a pumper.

WHEN the well had started paying, Harrington recommended to Lew drilling a second well on the Perryman land. It, too, was a very good well. So, they named the first well the "Perryman-1" and the second well the "Perryman-2." Then they completed a third well, the "Perryman-3," which was also very good.

Petroleum geology improved year by year as more and more wells were drilled in northeast Oklahoma. Harrington benefited from and contributed to the improvements. The improvements resulted in fewer dry holes.

With capital from the Perryman wells, and with Harrington's help, Lew drilled on other tracts that he owned. Most of these were good wells, which increased his capital. So, he started buying rights to drill on other land, again with Harrington's help. Some of the rights were short term; some were long term. And he further expanded his drilling to these new properties.

HIS DRILLING GOT to the point of squeezing his time for practicing law. So, he hired John Stevens to arrange and supervise his drilling, both with Harrington's help. Stevens had worked on a drilling rig, but he had to stop when he got injured.

Soon after hiring Stevens, Lew hired Harrington to work full time for him.

I've used his services so much that I might as well hire him. And I want to continue expanding my drilling, so I'll need his services even more than in the past.

Then he hired Ralph Mayfield, who had been working in a real-estate office, to acquire additional rights to drill.

With these hires, and with Pearl keeping the books on the operations, Lew freed up much needed time for practicing law.

STEVENS, Harrington, and Mayfield worked well together. They dramatically increased Lew's drilling and his oil revenues, which required hiring additional workers and acquiring equipment. With the expansion in workers and equipment, Lew concluded that the company needed a central facility for storing and repairing equipment and to serve as a headquarters where workers reported daily. He asked Pearl to work with Mayfield to find and acquire suitable land for the facility.

Pearl and Mayfield settled on a tract just beyond Sapulpa's city limits. It was centrally located and had good road access.

Once he had the land, Lew asked Pearl to work with Stevens, Harrington, and Mayfield to plan the facility and then engage a contractor to build it.

She finished the project in about four months. The men came to call it "The Yard."

Lew assigned an office in the Yard to Stevens. His duties had gradually expanded from all aspects of drilling to overseeing all employees and supervising the purchase and care of all equipment.

To give shape and stability to the operations, Lew created a corporation to conduct what had become an oil business. He gave ten percent of the stock to Stevens, ten percent to Harrington, and ten percent to Mayfield. In each case, the stock grant was in addition to salary.

WITHOUT SLOWING HIS PACE, Lew expanded the activities of the new company. It acquired drilling rigs and hired roughnecks to work on them so that it would not have to pay outsiders to drill. Also, from time to time, the company drilled for outsiders, which turned out to be profitable.

The company also developed the capacity to handle problems

with its producing wells. This required hiring and training roustabouts to do the work and acquiring the machinery, some of it quite large and expensive, needed for the jobs. But being able to do the company's maintenance was less expensive than engaging outsiders to do it.

IN THE MIDST of the depression, and because of the changes wrought by it, Lew saw business opportunities. People's lives and businesses had been severely shaken. There had been so many changes that business owners felt vulnerable. Many of them had come to want nothing more than stability and freedom from worry.

I want to acquire smaller companies in the oil business. I want to enlarge Riggins Oil Company and to integrate it vertically.

The idea energized him, and he aggressively pursued it. He did all the negotiations himself, leaving the details of deploying new workers and equipment to others. He offered cash, not company stock.

I don't want to dilute the present ownership percentages in the company.

When he was done, the company's payroll and equipment had increased by a third, and its daily production of oil had almost doubled. And it was finally vertically integrated. It could provide full service to oil businesses, large and small, from initial geology to caring for producing wells. It had become a formidable player in the state's oil industry.

22

SAM—SETTLING IN

It was winter. Sam worked on farming equipment; he kept up with daily chores: disposing of trash, hauling water from the river, and chopping wood for cooking and heating; and he hunted and fished.

He prepared rough plans for a house-improvement project and estimated the costs. Then he asked Lucy about it.

"Lucy, I want to improve the kitchen and bathroom. I want to build a work area and shelves in the kitchen. I also want to build shelves in the bathroom. I've estimated the costs and wonder if we have enough saved to cover them."

Lucy took a look at his estimates. "No. It would take everything we have, plus some."

"I hate to give up on the project. You've been working in the kitchen with planks on sawhorses. It would be more pleasant and efficient to work with a permanent table and shelves."

"I agree with that, and they would make the house look a lot better."

"What about borrowing the money? I've been thinking of using store credit again for seed. It might be less expensive to borrow that from a bank, and while I'm at it, borrow for the improvements also."

"I don't like the idea of borrowing. But I think we're going to have to, to finish the house. Why don't you talk to the bank about it?"

Sam reported back the next evening. "The bank is willing to loan to us. It wants us to say how many years we need to repay the loan. When it knows that, it will tell us the amount to pay each year. And it will require us to sign a paper giving it the right to sell our land, if we don't repay the loan. Whad'Ya think?"

"I don't like giving the bank a right to sell our land. Do you think we can talk them out of that?"

"I asked about that, and they said that was the only way they would make the loan."

"That's disappointing, but I know a family from the tribal community that had to borrow to build a house. They had to sign the same kind of paper. It's worked out okay for them. I suppose it's safe enough for us to borrow."

"How much time should we ask for?"

"How about two years? We shouldn't borrow more than we can pay back in two years

The next day, Sam and Lucy went to the bank. They signed a note and a mortgage. The bank made the loan in cash, which is what Sam wanted. After they left the bank, they went to a lumberyard where they selected the materials Sam wanted. Sam paid in cash, and the lumberyard said it would deliver the purchases the next day.

Sam finished the project without delay.

Lucy was very pleased with the improvements. Her kitchen chores became easier and more pleasant, and both the kitchen and the bathroom looked much nicer than before.

I feel good. I've done one of the things that I promised to do when we moved in.

It was time for Sam to plow and plant again. He went to the feed store and used more of the cash from the loan to buy seed.

About halfway into his plowing, his horse was not doing well. It had to stop and rest for a moment, several times a day. One morning when Sam went to the pasture to fetch the horse, he found that it had died.

"Lucy, we have a problem. The plow horse has died. I can't farm without a replacement. But it will cost more than we have. I think we need another loan from the bank."

"I agree. Why don't you check around and see how much a horse will cost?"

Sam found one. *That's a good horse, and I think it's a reasonable price.*

The bank had no problem with an additional loan, but it wanted the same paperwork as before. So, Sam took it home for him and Lucy to sign. By the next evening, Sam was back to plowing again.

In the middle of winter, Sam and Lucy noticed that his mother was not doing well. She was losing her color. She had trouble breathing. She had a dry cough. And she was sleeping more and more.

One morning, just before spring, she didn't wake up.

They buried her in the tribal community's cemetery. A few older people from the community joined in a brief ceremony. As Sam helped lower his mother's body into the grave, next to that of his father, he was overcome with emotion.

I feel smaller than before. I've lost another part of the

community where I grew up. Another part of me is gone. First the town. Then dad. Now mom.

I want to talk to Lucy about this, but I can't find the words.

———

A YEAR LATER, Sam and Lucy got an official notice from the state. It warned them to stop drinking water from the river, because agricultural byproducts had polluted it.

Lucy was concerned, Sam was skeptical.

"What will we do, Sam? We can't drink bad water."

"I doubt it's so bad that it'll hurt us."

"I don't agree. I insist that we do something."

"We could drill a well or I could haul water from another source. And right off, I can't think of another source that's anywhere near."

"Even if there were one, I don't want you hauling water every day any further than you always have. I think we need to drill a well."

"Fine with me. But we'll have to pay someone to drill it. And I'm sure we don't have enough money for that. So, we're back to borrowing more from the bank."

"Yes, I suppose so." And, sighing, "Oh, well. I suppose that someday we'll get out of debt and be able to start saving some money."

They borrowed more money and had the well drilled. It had a hand pump that was located about ten feet from the door. It was very convenient, and it relieved Sam of the daily chore of carrying water from the river. It also gave Lucy some piece of mind.

The well was expensive. Now, I'm more concerned than ever about our finances.

Lucy gradually noticed a change in Sam.

He seems anxious and is often more quiet than usual. *I'm worried about him.*
"Something's bothering you lately. What's going on?"
"Oh, nothing, really."
"Have I done something wrong? Are you mad at me?"
"No. Not at all."
"Good, that's a relief. But I wish you'd tell me what it is."
"I'm worried about our loans. They just keep increasing."
"I wondered if that was it, because I worry about them too. But it's not as if we haven't paid anything. We've been making a payment every year. So, I think we have to assume that everything will be okay."
"Yeah, but it'll be years until we're paid up."
"We can't think that way. We've got to look at what's right in front of us."
I'm glad she brought it up.
"That's a good idea. I'm going to try to think that way."
And he did. Every time he found himself thinking about the loans, he remembered Lucy's words. And in time he no longer thought about the debt. He put his energy elsewhere.

23

SAM—FIGHTING AGAIN

One evening a few weeks later, on the way home from buying supplies, Sam stopped at a small store that whites frequented. It had been a long day and he wanted to buy a pack of cigarettes. Several young boys, who apparently came together, were standing near the clerk, a pretty girl about the same age as the boys. When Sam was paying for his purchase, one of the boys, a tall, muscular kid, stepped towards Sam.

"Hey, Redskin, whatcha lookin' at?"

Sam ignored him, putting his change in his pocket.

"Hey, boy, I asked you a question. I expect you—"

Before he could finish his sentence, Sam landed an uppercut to his chin. The kid's knees buckled, and he dropped to the floor unconscious, bleeding from his muth.

As the other boys rushed to help their friend, Sam picked up the cigarettes he had purchased and walked out of the store.

I don't know whether I should tell Lucy about the incident. Probably, nothing will come of it. Then I would have worried her over nothing, and she would have gotten on me about fighting.

But the kid was bleeding. So, something might come of it, after all. I better tell her about it.

"Lucy, I'm home. How're you?"

"I'm doing well. I've had a good day. But you look like you're concerned about something. What's going on?"

"Something happened on the way here. I stopped to buy some cigarettes. While I was in the store, a young fellow insulted me. I hit him once, and he dropped like a sack of potatoes. As I left, I noticed that he was bleeding from the mouth. I'm afraid I hit him a little too hard."

"Oh, Sam. That doesn't sound good. What in the world did the boy say that angered you so much?"

"First, he called me, 'Redskin.' When I ignored him, he called me, 'boy.'"

"Sam, fighting has been getting you into trouble your whole life. I've told you over and over to ignore taunts from white people. You just won't listen. And now look at what you've done. I'm afraid that trouble's on the way."

Lucy was right. About a month later, Sam got a letter, which almost never happened. When Sam got home, she opened it, so that she could read it to him.

"It's too long for that. I'll read it to myself and then explain it to you."

Sam watched with growing despair as she read. She frequently sighed and shook her head, "no." And the more she read, the more agitated she became.

"Sam, you've really done it to us this time. The letter is about the incident when you were buying cigarettes. It's from a lawyer who represents the father of the boy. When the boy got home, his tongue was lacerated and two of his teeth were broken off. It will cost a lot to cap his teeth. And the lawyer thinks you should pay more than the costs, as punishment. He is demanding that you pay him more money than we'll ever have." She started to cry. "I don't know what we're going to do. I just don't know. It looks like we're going to lose everything we own, including our house."

I wish I had something to say, but I don't. I don't know how these things work out. But I intend to find out. I'm going to ask an elder about it. See if he can help me.

Lucy finally gave up and left him alone.

———

THE NEXT MORNING, Sam picked up the letter and went to see an elder who had worked with him in the tribal community and who knew him. Sam felt comfortable talking to him.

"Sir, I'd like some help, if you're willing."

"Of course, Sam. What's going on?"

Sam described the incident in the store. Then he gave the letter to the Elder, who read it through and then thought for a couple of minutes.

"Wow! You do need some help. Let's see if I can give it to you. Where do you want to start?"

"Can you explain what happens after a person gets a letter like this?"

"The process begins when the lawyer files a lawsuit against you. Then you'd need to hire a lawyer to defend you. Then your lawyer would prepare for a trial. He would need to investigate the facts. He'd probably try to talk to the other people who were in the store. And he would research the law. Then you and your lawyer and the other lawyer and his client would go to a court. Each lawyer would present facts and make arguments. Then a jury or the judge would make a decision on whether and how much you owe for what you did."

"Does it matter that the guy I hit started it by insulting me, calling me names?"

"No. Everyone involved in the process will be white. No one will care about what he said to you."

I don't understand what I'm hearing about the process, but I don't like the drift of it.

"Is there something I can do to prevent all this?"

"Yes. You can talk to the lawyer to see if the two of you can reach an agreement about what you will pay."

"Is there anything else I could do? I don't think I could talk to the lawyer."

"Let me think. Let's see. I could talk to the lawyer for you. Would you want me to do that?"

"Oh, yes. Yes. I'd really appreciate that."

"Well, I'll do it. But we need to decide what I'm going to say to him. Let's start with what you might be able to pay. What do you own?"

"That's pretty simple. One hundred twenty acres of farmland. Our house is on it. And some farm equipment. And a truck."

"Don't you have any savings?"

"No. In fact, I owe a lot of money to a bank."

"Does the bank have the right to sell the land, if you don't pay what you owe?"

"Yes, I think so. I don't really understand all the papers."

"Do you have any idea whether your loan is less or more than the value of the land?"

"By chance, yes. I haven't been able to make the regular payments that I'm supposed to make. So, one of the people at the bank recently told me that they could sell the land, but they would get less than I owe. So, they decided they would hold off, hoping I would be able to make some big payments and they would end up better off than they would if they sold it now."

"OK. I think I understand the situation. I'll go talk to the lawyer. Can you come back the day after tomorrow?"

"Sure. And thank you again. I feel a lot better about things."

SAM RETURNED, as requested.

"Hi, Sam. I think I've worked something out. You might not like it, but I don't see any other way for you to get out of this fix."

"What is it?"

"The lawyer sees that he can't get much if anything from your land. So, he'll let go of his client's claims against you, if you give the client a thousand dollars, to cover his son's dental bills. He's not happy with this, but he's willing to accept it."

"Do you think this is good for me?"

"Yes, I do. If you don't accept this, you'll have to hire a lawyer, which costs a lot. And if you lose, the other lawyer can force you to sell all your land, the whole hundred and twenty acres, and you might end up owing both the boy's father and the bank."

"Well, I trust your judgment. So, I'll go along. How do we finish it?"

"The other lawyer gave me a document that spells out the agreement. Just sign your name here, and I'll care of the rest."

After signing, Sam thanked the Elder and left for home, thinking about the situation as he drove.

I'm not looking forward to telling Lucy what I've done. She'll be relieved that the lawyer has been taken care of. But she'll be furious at losing money due to another fight.

Lucy was not angry. "You did well, Sam. Yes, you did well. I had expected worse. And we should be able to borrow the thousand dollars."

Great! I thought she would be furious.

24

LEW—DIVERSIFICATION

One day, Lew got word that Stevens wanted to talk to him. So, Lew drove to the Yard.

"What's going on, John?"

I'm surprised that he seems upset about something. He's usually unflappable.

"An Indian came by today looking for work. Wanted to be a roustabout. I told him we didn't hire Indians, because we've had bad experiences with them. They turned out to be troublemakers. I didn't tell him this, but the other men resented working with Indians. Hiring them had lowered morale and reduced productivity. Anyway, as he was leaving, he said that he wasn't sure you were the same person, but he thought you might have handled a law case for him. The more I thought about it, I decided that I should speak to you, to see if you wanted me to get him back in and offer him a job."

"What's his name?"

"Let's see, I have it here somewhere." After shuffling through papers on his desk, he said, "Oh. Here it is. Sam Perryman."

Lew thought for a couple of minutes. "No. I don't want you to change your policy. You're in charge of these men. You have responsibilities. If that's your policy, I trust that it's best for the company."

When Lew got home that evening, he mentioned the incident to Pearl. "I felt bad about it. After all, Sam's land has been very valuable to us. It gave us our start. But interfering with John's policy would have been bad business. It would have undercut his authority and made him afraid to make decisions. Besides, it's probably a good policy, if you judge it by Sam's behavior."

"I don't think you should feel bad about it. After all, you didn't tell Perryman to go to a saloon and start a fistfight. That was his own doing. And neither of you had any idea of the value of the land. And, as you say, business is business."

THE NEXT THREE years were busy for Lew.

But he never lost sight of what was going on at home.

He kept a close eye on Pearl to make sure she was okay and had everything she needed.

He occasionally had to forgo a party to which they had been invited, because of the press of business. Pearl was disappointed when this happened. She didn't complain, but she did have a good cry after removing her party dress.

Shortly after the Yard was completed, Pearl gave birth to a boy. They named him, "Mark." Lew was excited and couldn't do enough to help out. When he was at home, he changed diapers and helped Pearl with housekeeping. He painted a room blue that he and Pearl had designated for the newborn. And he personally bought baby furniture, most of which had to be ordered.

And Lew and Pearl kept up with Lew's family in Iowa. They

frequently corresponded with them. Occasionally, they travelled there to visit. At Christmas, Pearl sent them a substantial check.

LEW HAD A GROWING FINANCIAL CONCERN. *My wealth is accumulating so quickly that I'm concerned about my lack of diversity.* He brought this up one evening with Pearl.

"I have the same concern. Everything we have is in the oil business. It would be good to invest in something else."

"I've been thinking of banks. With the oil boom, several new banks have appeared. Some have been doing business for a couple of years. I know the people running them. They all seem sensible."

"Why don't you ask for financial statements? Tell them that we are thinking of investing. I can analyze the statements from the standpoint of balance-sheet strength and growth potential."

"Great idea!"

"How many you would want to invest in?"

"I've been thinking two. And I think we should try to acquire a significant percent of the stock in each one. If we're getting in on the ground floor, we might as well try for large gains over the long haul. And a significant amount of stock would make it more likely that we can get information when we want it."

"That's a good idea. And I suggest you get information on four of them. We can pick two from those four."

In the next couple of weeks, Lew got the financial information and Pearl analyzed it.

She recommended two of the four.

Lew made the purchases.

Pearl still had concerns about being too concentrated in oil. "Lew, I think we should start investing in land. It would give us additional diversification."

"That's an interesting idea. What type of land?"

"I think rural land would be best. Farmland or ranch land. It's less expensive than urban land, and urban land is tricky. Its value can be affected by all sorts of things—change in residential neighborhoods, city expansion, industrial development, to name a few. And tax bills are higher for urban land than rural land."

"That makes sense. What would we do with land that we acquire?"

"Farm it. Or run a ranch on it. We would probably net some income from it."

"I like the idea, but I doubt that I have the time to act on it. Would you be willing to handle it?"

"I would. But I would like Mayfield to work with me. We worked well together on the land for the Yard."

"I'm sure he would be happy to."

"And I will need to hire help to babysit Mark when I leave the house."

So, Pearl and Mayfield started buying land. And Stevens took on the job of starting and operating farms and ranches on it.

The Depression also produced many foreclosures of rural land. While Lew and Pearl were not happy about this, they saw no reason to refrain from acquiring more land at reduced prices. So, Pearl and Mayfield significantly increased Lew's land holdings. Over time, Lew became the owner of very substantial amounts of land.

The Depression did not affect Lew and Pearl personally,

except for their constant awareness of others' suffering. They donated heavily to hunger relief efforts.

LEW HAD ANOTHER CONCERN.

I'm over committed. With my law practice, and with my supervising the oil company with all its operations, and with my oversight of my land and bank investments, it's too much.

"Pearl. I've reached a decision that I want to discuss with you. I think I want to get out of my law practice."

"Why is that? You've always enjoyed it, and you're good at it. You have a steady stream of new clients, a testament to your results."

"With the growth of my businesses and investments, I just don't have the time to practice law. And I can make more money with the businesses and investments than I can practicing law."

"I understand you. But how can you get out of your practice? You can't just resign from all your cases."

"I could, by referring clients to other lawyers. But I wouldn't do that. I'm thinking I'll stop taking new cases. Then I'll finish the ones I have. It would be a way to gradually wind down and then stop."

"It's up to you. I hope you won't find that you regret doing it. But I doubt that you will, because you have always enjoyed your business and investment ventures."

25

SAM–A NEW PERRYMAN

Lucy doesn't look right this morning.
"You don't look so well. Are you sick?"
"No, not in the way you're thinking."
"I don't understand."
"Sam, I have some news for you. I'm going to have a baby. It's normal for a woman to feel bad in the morning when she starts carrying a baby. It'll pass in time. Until it does, I just have to put up with it."
Wow! This is a big change.
He hugged her. "Should you be resting? I can do what needs to be done in the house. I can even cook."
"Oh, no. I'll be fine. I don't need any more rest than usual. And having things to do around the house will help me forget that I feel a little nauseous. "
I'm not sure what to make of the news, how to react to it. I know that it's a natural result of a marital bed. But I don't know how it will affect our lives. Is it good news or bad?
He hugged her again. "I'm happy."

I'm not sure he meant it. He seems confused about having a baby.

While I'm not sure of my feelings about the baby, I know that we'll need a crib, and later, a child's bed.

He used his carpentry skills to convert his in-laws' bed into a bed for a child. And he built a mechanism to use part of it as a crib, while the child was younger.

JAKE WAS BORN in 1922 to Sam and Lucy.

Over the years, he turned out to be a blessing.

He was quiet, soft spoken, shy, and taciturn. He was not demanding.

He helped Lucy around the house. He wanted to help Sam with farming, which Lucy decried as too dangerous. But Sam let him help with house repairs and maintenance.

He was able to make friends, especially with other boys.

When he was old enough to go to school, he did well, even in the lower grades. By the time he was in high school, he had developed attitudes and habits that enabled him to do well with more difficult challenges.

He was a natural at athletics, participating in a number of sports, with which Sam helped him. When he got to high school, he was a benefit to the football team.

The school was so small that all able-bodied males were expected to try out for football, which he did. And he made the team, in a lineman position. The coach was impressed by his efforts, always giving the game his best. And he never got down or gave up when they lost, which they usually did because other schools in their division were larger and had a bigger pool of potential players from which to choose.

Wanting to make the best of his opportunity for a formal education, an opportunity that his father and mother had not gotten, he was a serious student. He worked hard at his courses and made good grades. He wasn't interested in social activities, drinking, or "hanging out." Nor was he interested in the school's extracurricular activities. He preferred to go home and do his homework.

He graduated at the top of his class. He was honored for his achievement, but this only bothered him, because he was very self-effacing.

I want always to do my best. Whatever comes of itis reard enough.

Sam and Lucy were continually proud of him and let him know it.

26

SAM–FARMING ON A LARGER SCALE

Sam evaluated his farming.
I have too much land to farm by myself. I think I'd come out ahead and be able to pay off more of our debt, by getting help. I don't want to hire someone. I don't want to be burdened by a salary. Instead, I'm going to look for a sharecropper. I'll offer him the use of 60 of the 120 acres. I'll use the other 60.

He found someone who would do it for two-thirds of the cash proceeds from selling his crop. But the sharecropper didn't work enough hours to make a success of it, and when there, he didn't work hard. As a result, the proceeds from selling his crop were small, and Sam's take from the arrangement was not significant.

The next year, Sam found someone else who would do it on the same terms. This person worked long and hard and had a good crop. But he didn't give Sam his share of the sale proceeds. Sam confronted him, but to no avail. He said that he was overdue on a loan and had to give all the sale proceeds to his lender.

Sam again visited an elder from the tribal community. "You

know about these things. Is there any way I can get what I'm due?"

"Is he an Indian or is he white?"

"He's white."

The elder thought for a minute or two. "Then, as a practical matter, the answer is 'No.' You'd have to hire a lawyer. And everyone at the court would be white. You would walk away with nothing."

———

SAM REEVALUATED what he was doing.

I'm going to try something other than sharecropping. I've heard about and seen farm tractors. If I could get one, it would enable me to efficiently farm all my land, without help.

Over the next winter, Sam looked for a used tractor.

He found someone who had decided to stop farming and was willing to sell his tractor, which was just two years old, along with his disc.

He approached Lucy.

She had her doubts. "The tractor's expensive, Sam. We have no savings at all, so we would have to take out a new loan for the whole price. We still haven't paid off our old loans. I'm afraid we'll never get out of debt."

"Over time it'll pay for itself. With it, I'll be able to farm the whole place. I'll get much more money from selling crops than I have in the past."

"What you say makes sense. I just wish you could find a way to use all the land without borrowing more money."

"You know I tried that. People don't want to rent land, because they don't know how much they'll earn on it. After a lot of looking, I found a sharecropper. He didn't do well. Then I found a second one, who cheated me. I'm tired of that solution."

"Well, OK. I'll sign the papers. But I hope you're right about doing better. I'd like to see us pay off some of the loans."

The bank agreed, and Sam soon had a tractor and disc.

THE TRACTOR *and disc will increase my enjoyment from my work. I'm doing something new.*

He threw himself into it.

It took him a while to get the hang of his new equipment, but once he did, he got the cropland prepared and seeded much earlier in the season than usual, even though he was covering more ground than before. And he got a head start on tending his crops.

As a result of getting an earlier start, along with favorable weather, he had a very good crop that fall.

Sam sized up the harvested crop. It was much larger than ever before.

I need a larger truck to carry this crop to buyers.

THE USUAL DISCUSSION with Lucy ensued, with the usual result. He found a used truck for a good price. The bank agreed to another loan. He was able to quickly and efficiently get his crop sold. And he made a larger than usual payment to the bank, which somewhat mollified Lucy.

The next spring, Sam had an unanticipated problem. The truck had left parallel ruts from picking up the crop. The ruts held water and the road soon became impassible.

I need a blade for the tractor, to maintain the road.

Another discussion with Lucy.

Another loan from the bank.

AND SO, it went with things that they found necessary to have.

The inevitable result was substantial debt on the land. But, as the population in the county had increased, so had the demand for good farmland. With increased demand came increased value. So over time, the value of the land came to exceed the debt. As a result, they had been able to buy a lot of things.

By modernizing, Sam had vastly improved his farming techniques. He had increased his yields and improved the quality of his crops. He was starting to chip away at his debt.

By necessity, these results depended on the weather. And, overall, the weather had held up. Some years were better than others, but by and large the weather was good.

That is, until 1930.

27
DROUGHT AND THE GREAT DEPRESSION

Oklahoma, including Sam's land, was abnormally dry in 1930. Sam had used precious cash to buy seed. But his crop was minimal and of low quality for want of moisture.

Moreover, the Great Depression was fully under way by harvest time. Prices for farm products had fallen substantially.

My sale proceeds didn't even cover the cost of seed. I'm disappointed. But all I can do is wait for better weather.

Unless belter weather's predicted, I don't want to spend the money to buy seed.

But the next year was no better. Nor for that matter, were the next eight years. Drought griped much of the Midwest. By 1934 the area had become known as the Dust Bowl. And the depression continued all the while.

Conditions were so bad that many Oklahoma farmers abandoned what they couldn't sell and moved to California. So also did farmers in Texas and as far north as Nebraska. People, particularly in California, came to call the migrants, "Okies."

IN THE ABSENCE OF FARMING, Sam had to seek employment, which was new to him.

He got a job with a construction company right away, because his reputation as a good carpenter preceded him.

I'm lucky. I got a job that I like without having to wait.

The construction company had a number of employees.

Working alongside others is different than farming. When I farmed, I was working by myself and for myself. So, I didn't think about what was my part and what parts were for others to do. I just worked as much as I could.

When he started working alongside others, he reverted to a mindset that he had developed at the tribal community. He made sure that he did his part by doing more than his part.

The company liked his attitude, as well as the quality of his work. So, he survived several job cuts, as the company finished the contracts that it had gotten before the depression.

But in the end, the company failed. Construction had simply come to a halt because of the economy.

HIS MINDSET FIT well into a depression economy. Other employers also liked his attitude. In jobs where he worked as part of a crew, and where there were job cuts, he was let go later than the other workers.

Sam heard that the oil business was surviving the depression, so he applied for a roustabout position at the nearest oil company. He was told that they didn't hire Indians. He applied at other companies, but to no avail.

So, Sam got in the habit of looking for work everywhere. He had a variety of jobs. He clerked in stores. He did odd carpentry and painting jobs. He was a night watchman at idle facilities. He did maintenance in commercial buildings. He performed janitorial services for retailers. He maintained track for the railroad. And he worked on a production line in a factory.

Sam experienced prejudice in the jobs market generally, not just with oil companies.

He spoke to Lucy about it.

"Whites get better jobs and higher pay than I can get. It isn't fair."

"Yes. But you're getting work. People are hiring you. That's better than your experience at oil companies."

HE HAD PERIODS OF UNEMPLOYMENT, which drove him to distraction.

I feel vulnerable. I've been pushed into financial straits by forces that I can't control, and, in the case of the depression, that I can't understand. I'm worried about the future. What will become of me, Lucy, and Jake?

He once got so down about being out of work that he spoke to Lucy about it.

I don't like talking to her about my problems, because I don't want to worry her, but I can't help myself this time.

"I can't stand it when I'm out of work. I can't stop thinking about it. All I can think about is how much I need work."

"I understand. I don't blame you. This is a scary time."

"The job I lost yesterday lasted only a week. It gets tiring looking for the next one."

"Of course. And I sympathize with you. But the only thing to do is to start again."

"Well, that's what I'm going to do. And, from experience, I know I'll feel better when I do. But I still worry about how we'll make it."

"Try to keep that out of your mind. Trust the future, so that you can put all your energy into doing what it takes to get another job."

And he kept at it. He never gave up. If one job ended, he started looking the next day for another one.

HE OCCASIONALLY REMINDED Lucy about their debt.

"Shouldn't we be making larger payments to the bank?"

"No. I want to make regular, but minimal, payments to the bank. We're living in uncertain times. We need all the cash we can get our hands on. Besides, we can't pay the whole debt, just small amounts. If we can't get out of debt, we're better off keeping our cash for emergencies."

He let her know when he needed cash to buy gasoline for his truck. But, aside from gasoline, occasional purchases of food, and small, irregular payments to the bank, she kept a tight rein on their savings.

She didn't have a bank account. She didn't trust banks, and some were failing. Instead, she hid their cash at home.

I know that Sam is holding back some of his pay, to buy cigarettes. But I'm not going to say anything about it. He's under a lot of stress. He's disappointed in not farming, He's anxious about getting work. And he's worried about our finances. He needs an outlet for his stress.

NO ONE KNEW how long the drought and depression would last. But Lucy's nest egg, coupled with her financial supervision and Sam's perseverance, had carried them. They had lived very frugally. They had lived with extreme anxiety. But they had not lost everything. They had a house and food, which many people didn't have. And while they had debt, they still had their farmland.

They made it through.

28
JAKE AND SAM—WORKING TOGETHER

It was 1939 and both the drought and the depression had ended.

Jake asked his father, " How about getting back into farming?"

"I'd love to do that. But I wonder if I would just be raising crops for the bank. I've tried, but I can't figure out what it would take to pay off the loan and end up with something for us."

"Maybe I can help you with that. Now that I've graduated, I want to work full time with you on your farm. I can start by figuring out what it would take to pay off the loan and whether it's feasible. How does that sound?"

"It sounds great. You and I can be partners. I would like that. But I don't want you to be worrying about whether we can pay off the loan. If we relax about it, all will turn out okay."

I am worried about it. Dad doesn't understand the situation, or he is being unrealistic. Or maybe both.

He visited the bank and asked to speak to a loan officer.

The officer was patient and explained things in detail. If Jake didn't understand something, he said so and the officer kept at it until Jake understood.

"The loan is what we call 'underperforming.' Your father's payments have usually been less than the annual amount due. And there were almost no payments during the depression. So, he is behind on the principal. Everything comes due in three years."

Jake was shocked. "I didn't think it was that bad. What's going to happen? Will the bank foreclose?"

"I don't think so. Not in the near future. The loan committee has been carrying the loan for several reasons. Your father has been making regular payments, which is more than some of our borrowers have been doing. We have a lot of underperforming loans. Right now, we're trying to avoid foreclosures. We don't want to end up with a lot of real estate. From what we can see, values are increasing. We think the value of your father's land will cover the remaining balance."

Jake took a deep breath. "Thanks. That's a relief."

"I do have a question. Your mother and father secured the debt with a mortgage on 120 acres. Because allotments were 160 acres, I have wondered about the other 40 acres. Do you know anything about them?"

Jake thought for a minute. "No. I don't know of any other land. I've never heard Mom or Dad speak about other land."

"I was just curious."

"I could ask Dad about it. But he doesn't like to talk about his affairs."

"No. Don't do that. It wouldn't change anything."

"OK. I'll let it go. But I have another question: Do you have any suggestions about what we can do?"

The officer thought for a minute. "You need a budget and a sinking fund. And you probably need to increase revenues by improving yields and quality."

JAKE THANKED him and drove directly to his school. He knew that the teachers were meeting to assess the year just ended. He found his math teacher, Mr. Blankenship, and asked if they could talk. He said he had a meeting but could talk afterward if Jake could wait.

After an hour or so, Mr. Blankenship emerged from the meeting and they went to his office.

"What can I help you with?"

"Please explain how to set up a budget and a sinking fund."

Jake described his meeting with the bank officer. "The farm is all my dad has. I want to help him save it."

Mr. Blankenship spent several hours teaching Jake how to create a budget and a sinking fund. When they were finished, he leaned back in his chair and sighed. "You know, the banker said something else—that you need to increase revenues. I can't help you with that, but your Ag teacher can. Why don't you look him up? He knows a lot about farming."

The next day, Mr. James, the Ag teacher, also spent several hours with Jake. He explained the importance of high-quality seed, the need for fertilizer and herbicides and when to apply them, when to harvest, and where to sell crops.

Jake thought about it that night.

I need cost and revenue estimates to do a budget.

The next day he went to the feed store. He asked about different types of seed, fertilizer, and herbicide. He asked what the owner would recommend. He asked about quantities needed. He asked about prices. And he kept copious notes.

The son of a nearby farmer had played on Jake's football team. So, Jake felt comfortable asking the farmer for help. Jake went to his house and asked about estimating yields and prices.

The farmer respected Jake and what he was trying to do, so he spent an afternoon helping him. Again, Jake took notes.

Jake and Sam worked every day on the land, preparing it for planting.

For several nights, Jake worked on a budget. He wanted to involve his father, but his father resisted. Sam finally sat down with his father, but he fidgeted.

"Why spend time on a budget? Let's just get going and see what happens."

Jake was not dismayed.

"But, dad, we have to come up with something to guide our purchases and sales, so that we have a chance of paying off the loan."

"Oh, son. Don't worry about it. In the good years before the drought and depression, I made payments every year. If we show them now that we're trying, and if we make any payments, they won't shut us down.".''

I don't want to tell him what the loan officer said. It would just discourage him. He's doing half the labor. I want that to continue.

So, Jake kept working on the budget by himself and finally arrived at one. Early the next morning, he took it to the bank and asked to speak to the same loan officer.

"I'm impressed. You've worked hard and learned a lot in a short time. And your budget looks realistic. But look at your sinking fund. It will pay more than in past years, so that you can make some

inroads on the principal. But not enough. I doubt that you can do any better than your budget. And if you make budget this year and the following two, you still won't be able to pay the loan by its due date." Jake had concluded this from his calculations and had thought about it. "Would the bank consider extending the loan?" The officer thought for a minute. "It wouldn't now. Your father has been good about making regular payments, but they haven't been large enough. However, if you do well and increase your payments for a couple of years, it might change things. But, to be honest, you probably wouldn't get an extension."

Jake was set back a bit, but he kept calm. "I appreciate your being honest with me. As you can tell, I'm new to this, and I'm trying to learn a lot in a short time."

"I know, and I don't want to disappoint you. But I don't want to make empty promises either. I want you to have good information to guide you in deciding what you're going to do. I hope you understand."

"Yes, sir. I do. And I thank you for your time."

Jake made up his mind before he reached the door.

I'm determined to do my best, and that doesn't include giving up.

JAKE DIDN'T WASTE any time. He started right away, by buying seed. They were short of cash, so he negotiated credit at the feed store, showing the budget he had made. The owner decided to take a chance on Jake. He saw the budget, and he knew about Jake's reputation in the community, based on his performance during high school. So, they agreed on a rate of interest and agreed that Sam would pay the account at harvest, before he paid anyone else.

Jake explained the agreement to Sam.

I'm happy to let Jake make the decisions and do the purchasing and selling. I don't have the knowledge that he has picked up from his teachers.

Jake quickly saw that working on the farm now would be different than working on it when he was in school . Now, *he was on fire.*

I understand what needs to be done, and I aim to do it.

He worked from sunup to sundown every day. And he felt great about it.

Sam more than once mentioned the difference Jake was making. "We're getting much more done this year."

At harvest time, they could clearly see the difference. The farm had done better than ever. After Jake had settled his account at the feed store, he made a payment to the bank. He was a little short of the amount due for the year, but it was the largest payment to date on the loan.

Sam was impressed. "I hope we can repeat the performance next year. We had a lot of luck this year. We got a lot of rain. Insects weren't that bad. Prices were high. Let's hope we have the same luck next year."

Nevertheless, it was fall and he was ready for a rest from the hard work of the summer.

But Jake was just getting started. He had plans for the winter. They would repair and clean tools and equipment for a good start next summer.

They worked in the barn. Unlike working in the fields, this gave them an opportunity to talk. So, Jake learned a lot about his father that he had not known before.

Jake thought, *this is a very worthwhile benefit of working together.*

They did not have long days. As daylight grew shorter, Jake was able to work after dinner on the budget for the following year. He started fresh, getting information about projected prices for supplies and crops. When he was finished, his sinking fund would permit him to make a full annual payment on the loan.

It's an ambitious budget, but I'm determined to stay within it.

The second year was better than the first. Everything went well. Luck was with them, and Jake made a full annual payment on the loan.

BUT THE THIRD year was different. The cost of supplies had risen a good bit. And harvest prices fell because of bumper crops in the Midwest. As a result, Jake's bank payment would be the smallest in the three years during which he had participated.

I'm disappointed and concerned. I don't look forward to visiting the bank, but I have to do it. I have to ask for an extension.

So, he asked for the same loan officer, when he went in to make the payment.

"Good morning, sir. I've brought in this year's payment, and I have a request." Jake had to take a deep breath. "You remember when we talked about an extension a couple of years ago? You said maybe, if I did well for the next three years." Jake had to breathe deeply again. "I think I've done pretty well. Not perfect. But pretty good. My lowest payment was larger than any payments before I took over. I think that I can get the loan paid if I have more time."

The officer was uncomfortable. "I was afraid you would ask that, because I have bad news for you. The loan committee reviewed the loan just a week ago.

Everyone was impressed and pleased with what you have

done. But the loan has been underperforming for too long. We decided, reluctantly, that we had to call it.".."

"So there's really nothing that can be done?"

"I'm afraid not."

Jake had to fight back an urge to cry. "So, dad's going to lose his farm?"

"I'm afraid so."

I failed. I dread telling my father. I let my family down.

"I understand. We'll just have to live with it."

"It's not your fault. Probably not anyone's fault. We had terrible drought years. And the depression just kept going and going. It was a bad time for a farm loan. But no one could have foreseen that, when the loan was made. I hope you understand what I'm saying and take it to heart. You're young and have a bright future. And you made a valiant effort to salvage the farm. I'm sure your father is proud of you for that."

"Thanks. But it's going to take some time for me to get used to it."

"Sure it will, and I hope you're patient about it." He paused, collecting his thoughts. "In another vein, I have a suggestion. I think your father can sell the farm, holding out the house and an acre or two around it, for the amount of the loan balance, and maybe even a little more. That would avoid the fees and embarrassment of a foreclosure. Think about it. It might soften the blow."

"Thank you. I appreciate the suggestion and your concern."

Jake went home with a heavy heart.

I'm deeply disappointed. Many things haven't come easily for me. But I've usually gotten what I wanted, when I put my mind to it. But not this time.

It's like losing those high-school football games. Except that this is not a game. It means so much more.

But in both cases the important thing is what I put into them.

I put my mind to them and did my very best. I never got discouraged, and I never gave up.

JAKE EXPLAINED the situation to his mother and father. "I urge you to put the farm and farming equipment up for sale. I think it will be best for you."

A couple of days later, Lucy told Jake that they had decided to accept his suggestion. "But Jake, we need your help to do it. Your father would never admit this, but he has no idea how to go about it. Maybe the man at the bank could help you."

"Sure, mom, I'll do that."

So, he found a realtor to sell the land. And he found an equipment dealer to sell the tractor and blade.

When it was all done, his parents were free of debt, got to keep their home, and received a small amount of cash.

AS TIME PASSED, Jake came to view the three years as a learning experience.

Not everything works out well. I can't be hard on myself about that. What the banker said was true. There are a lot of things I can't control. I just have to accept them and go on with my life. All I can do is keep doing my best and maintain my optimism by assuming that things will work out well. But above all, I can never give up.

TULSA

Lew started thinking about a relocation.
I think it's time to move to Tulsa. It's much larger than Sapulpa. And it has more activity than Sapulpa.

"Pearl, what would you think about leaving Sapulpa?"

"I must confess that the thought has never occurred to me. But I think I'm open to the idea. Where to and why?"

"I'd like us to live in Tulsa. More and more oil businesses are locating their offices there, which would give me the opportunity to meet others in the business and possibly learn from them and even do business with them. And with the availability of automobiles and the development of a suitable road between Tulsa and Sapulpa, I could easily finish my few remaining law cases and stay in touch with the Yard."

"I see what you're saying. But what about the impact on Mark? He's familiar with his school here and has developed a lot of friendships."

"I've thought about that. He's still young, and as you say, he's good at making friends. It would be an adjustment, but I think he could handle it, particularly if we move in the summer.

And, based on some looking I've done, I think he'd have better schools in Tulsa, especially as he gets into higher grades."

"It's an interesting idea. But let's not be hasty. Let's think about it."

Pearl did her own investigation and concluded that the move would be for the best.

AFTER REACHING A FINAL DECISION, Lew and Pearl drove to Tulsa to look for a house. By chance, they wandered into a special situation. They found a mansion at the top of a hill that sloped to the banks of the Arkansas River. The lot was an entire city block. And the house was reasonably priced, considering its size and location. The owner of a successful business had just finished building the house and had died unexpectedly before he and his family could move in. His widow lost all interest in the house and was concerned about the debt incurred to build it. Lew and Pearl loved it and signed a contract to buy it the day they first saw it. Because of their accumulated wealth, they paid cash and did not have to borrow.

ONCE IN TULSA, Lew found space suitable for the company's headquarters in a downtown office building. It was a suite with five offices. Lew, Harrington, and Mayfield took three of them. They started working in their new offices without delay.

As Lew had predicted, the move to Tulsa was beneficial to his business. He wound down his law practice and devoted all his time to business. He seemed to have boundless energy, and the company continued to grow. He met additional people in the oil business and was invited to join a newly formed downtown club for oilmen, where he met people who were valuable connections for his company.

Lew and Pearl quickly became part of society. They gave elaborate parties in their home and received reciprocal invitations to other parties. As in Des Moines and Sapulpa, they became well liked and sought after. It wasn't long before they were invited to join a country club.

Through connections at the oilmen's club and at the country club, Lew became interested in politics.

I have no desire to run for office, but I want to be involved in politics. I want to contribute to candidates and participate in campaigns.

Because of his significant help, he came to know office holders, in both the Oklahoma legislature and in Congress. As a result, Oklahoma newspapers considered him to be influential and turned him into a public figure.

AS THE DEPRESSION was winding down, a group of men asked Lew to participate in creating a new country club. He accepted the offer and threw himself into it. At their initial meeting, they made decisions about the facilities, amenities, and services that the club would offer. They decided that everything would be first class. As a result, it would be very expensive to join the club. They also decided that the club would be very exclusive. Membership would be by invitation only and limited to people who were prominent in the Tulsa community. And club policies would be developed to assure that membership was prestigious.

Initial invitations went primarily to oilmen, who were community leaders by virtue of their wealth. Later, successful businessmen, politicians, and cultural leaders were also invited.

Lew resigned from his initial country club so that he could

focus his attention on the new one. And he devoted a lot of time to the new one. He took a major role in administration, utilizing his business skills. He materially helped the Club migrate from idea to reality.

I'm very proud of the Club. It's a shared accomplishment.

The Club grew quickly and flourished, ultimately becoming one of the most prestigious country clubs in the nation.

Because the Club had become an important part of his life, he remained a prominent member and Director until his death.

JAKE—THE ARMY

"Mom. Dad. I have something to tell you."
His parents took note of his expression and his tone of voice. They knew it was important. So, they stopped what they were doing, sat down, and by their silence let him know that they were paying attention.

"I've joined the Army. I report for basic training in a week."

His mother looked shocked and then, gradually, sad. "Why'd you go and do that?"

"I decided that it was my duty. You've seen all those signs saying, 'Your country needs you.' We're at war, and I don't want other people doing my fighting for me."

She started to cry. "I hate to see you go. I'll miss you and I'll worry about you every day."

"Aw, mom, come on. Don't cry. I'll be all right. I'm sure everything will work out well."

Sam was taking it all in. He was uncomfortable with crying. To change the mood, he looked at Jake and then paused a bit, collecting his thoughts. "You're an Indian, son. People will be watching you. Make sure you do your part, son. Don't leave it to the other guy to do."

"Don't worry, dad. I will."

BASIC TRAINING WAS NOT a big deal for Jake. He was already in good physical condition, so he was not bothered by the conditioning drills.

Most people complained about the food, but not Jake. The food's OK. It's simple and plain, which is what I'm used to. So, I have no cause to complain.

He applied himself as he had in school. He studied the manuals and paid attention to instructors, both in the field and in classrooms. As a result, he learned what he was being taught. And the instructors were impressed.

When he was done with boot camp, he was assigned to an important desk job at Fort Sill in Oklahoma. This was due to his performance in basic training. The Army needed someone to fill the job, even though the country was at war, and the decision makers thought he would be the best candidate.

SAM CLOCKED out at Midnight and walked to his truck, to return home. It was slick underfoot. He had to be careful walking. A winter storm was approaching. The leading edge was producing freezing rain, not uncommon in that part of the state.

He thought about his options.

I know I shouldn't drive, especially because my tires are old and worn. But I don't want to spend the night alone in a cold hotel room. I want to sleep in my own bed, with Lucy. So, I'll drive slowly.

But on a sharp curve the truck fishtailed. By the time he could get it under control, it had skidded off the road into the left margin, facing oncoming traffic.

He couldn't get it to move. The tires just spun on the ice.

He decided to wait a while in the truck to see if someone happened by. But no one did. At that time of night and with those road conditions, Sam was not surprised.

So, he decided to get out and see what he could do. It was dark. The cloud cover of the storm blocked any light from the sky, and he was far from any habitations. When he stepped out, he immediately lost his footing and slid down a steep embankment, striking his thigh against a sharp rock.

He was in terrible pain.

I've broken my leg.

He tried to stand but couldn't. So, he tried to pull himself up the embankment. But he couldn't get a grip on anything because of the ice.

A north wind was blowing, and it was very cold.

I need to think of alternatives.

But he couldn't focus his attention, because of the pain and because of discomfort from the weather.

Hypothermia began to set in. It became increasingly difficult to breath.

Realizing that the end was near, his thoughts ran to what he was leaving behind.

My sole legacy is a confusing and disappointing world where people treated me unkindly simply because I was an Indian, and where, try as I might, I was never able to accumulate anything to show for my efforts, let alone provide a comfortable life for myself and my family, and where, even though I was given a head start with 160 acres and some cash, in the end, I lost it all.

Finally, he drew his last breath.

JAKE WAS at Fort Sill when he got the news that his father had passed away. He requested and received leave to go home. It took most of a day on Greyhound buses.

I want to say "Goodbye" to my father.

I'm glad I'm here, for my mother's sake.

On top of being sad about Sam's passing, his mother was very worried about her financial future. She felt she could not talk about this with friends, so Jake was a welcome listener. And he assured her that she would be okay. He told her that he would have the Army send her his pay, and that he would add her as a joint owner on his savings account. He reminded her that she was frugal and that she had survived some financially tough times. Talking to Jake calmed her, and she was able to properly mourn her loss.

They buried Sam in the community's graveyard. The grave was next to those of Sam's parents. There was a traditional Creek Indian ceremony. A handful of people from the community attended. At an appropriate moment, Jake stepped forward and spoke. He was wearing his Army uniform.

"Dad spent his childhood in one world and the rest of his life in another world. His new world was quite different than his old world. He had problems and never understood that the source of them was the context into which he had been thrust.

"The tribal community was a simple society with few rules, all of which were intuitive.

"By contrast, the world outside the tribal community was complex and sophisticated. It ran on many rules, often complicated and non-intuitive.

"In his new world, he encountered uncontrollable setbacks. The first was discrimination. Later it was drought, and then the Great Depression.

"From the outset, he was saddled with debt. He got land when he left the community, but very little cash. He and mom couldn't use land to buy things they needed to set up a home. They had to borrow, using the land as collateral. And once they got into debt, they never got out of it.

"More important than his setbacks and problems, was the way he faced them. He never gave up. He kept doing his best. He did his part, and then some. And that's how I will remember him.

"Thanks for listening to me today. As I rode here to attend this ceremony, I thought about what I wanted to say. Participating here has helped me to understand Dad, to forgive his shortcomings, and to love him even more than before."

31

JAKE—D-DAY, THE ALLIED INVASION OF NORTHERN FRANCE ON JUNE 6, 1944

Soon after returning to Fort Sill, Jake got orders to report for a new assignment. He quickly learned that all the younger people in the office had received the same orders. They were told to pack combat fatigues and meet at the mess hall at 7 o'clock a.m.

The next morning, they took a bus to the train station. The train took them to Norfolk, Virginia. From there they boarded a naval troop carrier. A week later, they disembarked in Southern England.

AS SOON AS people arrived at a makeshift base, they began briefing sessions.

The Allies were assembling an expeditionary force to cross the Channel and invade the northern coast of France. The new arrivals studied aerial photographs and maps. And officers explained the purpose and mechanics of the invasion.

They would be ferried to a beach in amphibious landing crafts. The immediate objective was for each squad to cross its part of the beach, climb the hills that rose from it, and penetrate the Nazi lines behind the hills. The ultimate objective was for all the squads to regroup, assemble a battalion, and begin an assault against the German troops occupying Western Europe.

Ordnance issued ammunition and hand grenades to everyone who would be involved in the landing.

Jake approached the briefings in the same way that he had approached other new situations. He did his best to assimilate all the information, so that he could understand the operation and his role in it.

One evening, officers informed the members of the force that the beach landing would be at 0700. Jake's squad was told to muster at 0530 in the area where the Army had its busses, grouped there to take people to landing crafts.

On the bus to the landing craft, the Sargent in charge of Jake's squad spoke to the men.

"You've been briefed. You've seen the aerials. You know what to do—cross the beach, climb the hill, penetrate the German line." He paused, as he thought about how to say what came next. "We expect resistance, but we don't know how much and where. Almost certainly there will be machine gun installations on the hill. So, there's no use candy coating things. It's a dangerous mission."

He regarded his men with sadness. All of them were young and probably scared. "Good luck." He couldn't think of anything else to say.

They boarded an Army Higgins boat. It was 36 feet long and 10 feet wide. It held 36 men. Jake's squad filled about a third of it. Two other squads filled the rest. Jake and his squad were at the front, because they would disembark first. They were to deploy straight ahead. One of the other squads was to deploy to the right and the other to the left.

Once loaded, the boat started its engine and began racing toward the French coast.

Jake had heard that the Channel always had rough water.

I hope I don't get seasick.

While he saw many of the other men in the craft vomiting, it didn't happen to him, probably because of an adrenalin rush.

THE OVERCAST SKY WAS SURREAL. It was smoky from Allied offshore artillery fire and the exhaust of larger ships. It was crowded with Allied fighter planes and blimps. And there were frequent flashes from exploding artillery shells that had found their targets on shore.

There were ships of all sizes and descriptions—from large ships carrying heavy artillery to minesweepers and to a seemingly endless number of landing craft, ferrying thousands of fully equipped soldiers to shore.

There was a chronic blare—the engine noises of ships and landing craft, the droning of aircraft, the thunder of large artillery fire, the fierce explosions when their shells hit home in the distant hilltops, and the detonation of mines by the crew of a minesweeper. At times, it was almost unbearable.

They had sailed into and had become part of a hellish scene.

As they got closer to the beach, Jake began observing on a larger scale.

This is a huge and complex operation. The individual components are beginning to coalesce into a colossal fighting force. I've never seen anything like it. I'm impressed by how coordinated the components are and by the drama of what's unfolding. How can a man escape such a fury?

Jake had never been outside Oklahoma. Now he was participating in an invasion of France.

I have to avoid getting overwhelmed. I want to remember to do my best. And I want to remember Dad's admonition, "Do your part, son."

I need to relax and to focus on my specific job.

As the craft reached the shore and began to lower its front ramp, Jake saw a distant beach with a scattering of injured and dead soldiers. This brought home the danger of the mission on which he was about to embark.

He took a deep breath, secured his helmet with its chin strap, and jumped onto the beach.

HE WAS the first man off the craft. He immediately flattened himself against the sand, to reduce his profile. The rest of the men in the squad followed suit.

Jake took stock of the situation. The beach was full of soldiers advancing toward the hills. While he heard bursts of machine gun fire from the hills, none was near his squad.

He and the others got into a crouch and started slowly moving up the beach, rifles at the ready. They reached the hill without incident.

Jake again reconnoitered. He looked closely at the hill. Seeing no movement, he started climbing.

The bottom of the hill was about a foot above the level of the sand. So, Jake jumped up and started cautiously walking, in a crouch. The rest of the men hung back, remaining in the partial cover provided by the drop from the hill to the sand.

When Jake was halfway up the hill, machine guns erupted from ahead and above. Jake felt a searing pain in his right thigh.

The force of the bullet spun him to the ground, and he was momentarily stunned.

After clearing his head, he again took stock. The whole squad was on the ground. He couldn't tell how many had been hit. When he looked up the hill, he could see a machine-gun nest. It contained two tripod-mounted guns. It had been camouflaged with tree limbs, which the gunners had removed just before firing.

There was a steady flow of gun fire from the nest, but the bullets didn't come near him.

I must be too low to the ground for them to see me. Or else they're focusing on the other men in the squad.

"Do your part, son" echoed in his mind.

He considered what he could do. He knew that the gunners could wipe out the whole squad. Having applied himself diligently in basic training, he had learned well. He knew exactly what to do.

I need to get closer to the nest.

He left his rifle behind. It would make it more difficult to crawl up the hill.

His right leg was useless, requiring him to crawl up the hill, dragging the leg. This was very painful.

I'm much closer to the nest than the other guys in the squad. And time is a factor. I want to deal with the gunners before they hit more guys in the squad. And I want to do my very best. So, I'll ignore the pain and move quickly.

Being right-handed, and remembering his basic training, he aimed to get to the left of the nest and about fifteen feet away.

I'm finally in a good spot. It's time to act.

He removed his helmet, so it would not restrict his movements. He rolled over onto his back, although doing so was very painful. Then he removed a grenade from his belt, put a finger through the ring, pulled out the pin, let go of the release bar, counted, and lobed it into the nest.

The last thing Jake saw was a blinding flash of light. The last thing he heard was a deafening explosion. The last thing he felt was a violent land swell that rolled him onto his stomach and coated him with freshly dug dirt.

32

MARK—LIVING IN HIS FATHER'S SHADOW

Mark was born in 1917 to Lew and Pearl. His childhood was easy. He never experienced hardship. He never wanted for anything. He had no work to do. But he was not spoiled. His mother made certain of that.

Some of his earliest memories were about Lew's first country club. He practically lived there. His parents were busy, and the club was a convenient and acceptable form of babysitting. One of his parents' domestics drove him there in the morning and picked him up that evening.

When school was in session, and when he was in the lower grades, he spent Saturdays and Sundays at the club. He also went there after school, if he had something special, like a tennis lesson.

He learned to swim at a young age. As soon as he could hold a racket, he started tennis. Not long after, he began golf. He had professional instruction with all three.

He socialized easily. He got to know the other children at the club, got along with them, and they liked him.

In the lower grades of school, many of the other children lacked his social skills and polish. But he got to know them and got along with them. They felt at ease with him, notwithstanding that they sensed that he was different than they.

When Lew resigned his first country club, and helped start his second one, Mark transitioned easily to the new one. Its programs for children were practically the same as at the first club, except that the facilities were newer and nicer.

———

ALTHOUGH HE DIDN'T REALIZE it, in the upper grades, especially high school, his last name opened doors for him. People seemed to like him before they met him. Especially girls. They paid attention to him, even though they were standoffish with many of the other boys.

Other boys noticed this. Most of them admired him for his attractiveness to girls, rather than being jealous about it. They tried to emulate him, although they made pains to avoid being obvious about it.

He was easily the most popular person at the school. He was elected President of the senior class. At the beginning of home football games, he led the team from the locker room to the playing field. And he chaired the prom committee.

He attended parties and dances hosted by parents of students. And he went on dates with fellow students. Unlike many of the boys, he never had any trouble finding a date.

———

I'M MORE interested in socializing than in my studies. Father and mother do a lot of socializing. And father seems to know every-

one. So, I see the value of socializing. But I don't see the value of high grades. With father's business connections, I'll always be able to get a job.

He was a below-average student. He had no motivation to do well, so he didn't respond to the challenges of his courses. By and large he neglected homework assignments. He did just enough to get by.

"C" is good enough. I just want to make sure it Isn't an "F". Father and Mother would have a fit if I got an "F" in a course.

Mark thought about trying out for the varsity football team.

Football players seem to have everyone's approval. I'd like to be one of them. I'm good at tennis and swimming. I'm a pretty good athlete. I could probably make the team.

But I don't want to try out for it. It's a rough sport, and I don't want to get hurt.

HE WAS NEVER uncomfortable at social events. The Club taught ballroom dancing and manners. A Club member, Edith, ran the program. Mark found her easy to talk to. Somehow, he found it difficult to talk to his parents about important questions. So, when he had an issue, he always spoke to her.

"Edith, do you have a minute?"

"Of course, Mark. What's going on?"

"There are a few guys at school who're rude to me. They keep saying that I'm part of the 'country-club set', and they make ugly statements like 'What're you gonna do with all your money when your father dies?' I don't know how to deal with them."

"You're going to encounter this all your life. Some people will be jealous of you. You have to rise above them. Ignore them.

If they don't get a rise out of you, they'll stop bothering you. But if you talk back to them, they'll never leave you alone. So just walk away."

Mark always took her advice seriously. So, it wasn't long before the boys stopped bothering him.

HIS POPULARITY DIDN'T GO to his head. It wasn't even important to him.

What I really want is approval, not popularity. Other kids seem to like me, and at times even play up to me, but I sense that they don't approve of me.

Mark had never gotten approval, or much attention at all for that matter, from his father, who had always been too busy to notice his son's emotional needs. The result was the beginnings of existential shame, which leads to an excessive need for attention and approval.

MARK WAS NOT an instant success at college, unlike high school.

He chose to attend Oklahoma University in Norman, Oklahoma. Norman was a little more than 100 miles from Tulsa, farther than the reach of Lew's reputation. Neither "Lew Riggins" nor "Riggins Oil Company" was a familiar name there.

For the first time in my life, I'm on my own. I'm not getting the attention that I got in high school simply by saying, "Hello" to someone.

The lack of easy attention shook his confidence.

I wonder if something is wrong with me?

He couldn't see that the problem was not him, but the

absence of his father's good name. It was only later that he realized the importance of a last name, particularly his last name.

I feel alone, even though I'm surrounded by people. I feel ignored, inadequate.

But because of his social skills, and because his clothing and personal care befitted someone who was "raised right," he was invited to join a fraternity. That gave him some relief.

Over time, he made friends, both in the fraternity and outside it. But they were surface friendships only. He didn't talk to his friends about his feelings. He didn't know how. He had never observed his parents' discussing their feelings. And his high school relationships were not close enough to include talk about feelings.

He gradually fell into an emotional habit of withdrawal and isolation.

———

BUT HE NEVER STOPPED ENGAGING. He participated in sports—golf, tennis, and swimming, earning a position on the varsity teams for all three. Also, he was always up for a card game. And he was active in fraternity affairs, often volunteering to help with planning social events.

———

NOT SURPRISINGLY, given his poor grade performance in high school, and his lack of interest in his studies, he did poorly in his courses.

He kept his grades to himself. But most students talked about their grades.

I can't help but notice that my friends do better than I. When I compare myself to them, I feel inadequate.

I've decided that I just don't have what it takes to make good grades. When I think about it, this causes me to feel bad about myself. So, I'm just not going to think about grades. They don't offer me a chance for approval.

While he didn't realize it, he was always looking for approval. When it came, he doubted or discounted it. It was never strong enough to be convincing.

EARLY IN HIS FRESHMAN YEAR, he was introduced to alcohol. His parents had not given him alcohol, saying that it was for grownups. Sponsors of high-school social events had not permitted drinking. And the Club didn't serve minors.

So, when he was offered a beer, he was anxious to try it.

Even though this is my first, I can tell that it's a miracle worker. I feel like I've found a loyal friend. I feel okay about myself, for once.

That beer was just a starting point. He was soon having two or three. He was always ready to drink a couple beers if someone in the fraternity house wanted company. And there was often someone who didn't want to drink alone.

Friday and Saturday nights came to be a routine. He always drank beer, alone or with others.

Drinking had become a way of momentarily forgetting himself. When he drank, he didn't think about grades and didn't compare himself with others. He didn't feel bad about himself. He felt that he fit in, measured up.

Fortunately, he was non-confrontational. Intoxication did not lead to belligerence, but rather to clowning and other ploys to please people. When the fraternity had an event, he sometimes played the role of "life of the party."

At a fraternity party midway in Mark's junior year, he

couldn't stop drinking. He just drank until he passed out. Amidst of a lot of gawking, fraternity brothers carried him to his bed.

When he awoke the next morning, he felt terrible. He had a headache and was nauseated. He stayed in bed all day, finally getting up for dinner. When his head cleared, he thought about what had happened.

I was just careless about how much I was drinking.

He didn't realize the implications of the episode. Nor did he see the significance of the fact that he couldn't remember anything that had happened after a point early in the evening.

By the middle of his junior year, he had earned the reputation of fraternity drunk.

I hope it's good-natured ribbing. And I also hope it doesn't get back to my parents.

Unfortunately for Mark, it did get back to his parents.

AFTER HIS JUNIOR YEAR, he decided not to return to college.

I've failed so many courses that I can't graduate after four years. There is no way to hide that, but if I drop out, no one will know why. This will avoid disgrace for my parents. And I can always tell people that I decided that I wouldn't get much more from a fourth year than I had already gotten in three years.

He didn't volunteer to explain his decision to his parents. And they didn't ask him to explain it. They knew the reason, and he knew that they knew.

AS MARK WAS DRIVING HOME, he reflected about the impact of his three college years.

Father has a good self-image. By contrast, I've arrived at a bad self-image.

The comparison shocked him; he so wanted to be like his father. The realization almost destroyed what remained of his self-esteem.

All in all, college had been a rough go for Mark.

33

BACK HOME

For a couple of months, Mark visited high-school friends and played a lot of golf and tennis at the Club. But living back at home, every day he sensed his parents' disapproval. He did some thinking.

This is not, after all, a summer break. It's the beginning of a new phase in my life. Further education is out of the question. The only option is work.

He let his parents know when he started job hunting, and that relieved the tension in their home. But he didn't tell them where he was looking.

I don't want to interview people who know my parents. I don't want them to know that I haven't graduated from college. Because father has a high profile in the business community, and because mother is socially prominent, there aren't many "suitable" options left.

He visited several out-of-the-way retailers. Each person he interviewed identified his family once they noted his appearance and saw his name. And they were not impressed. To the contrary, they had preconceived notions about him. He heard consistently that he had no useful skills.

"Everyone in Tulsa knows that you haven't worked a day in your life."

These interviews were hard on Mark. After several of them, he went out wandering, not wanting to come home, and wanting a beer.

I'm not used to the candor that I'm encountering. Other people I've dealt with care about me and take the time to be diplomatic.

The experiences did nothing to help his battered self-esteem, including feelings of inadequacy.

Father's reputation has been a negative factor in my job search. I'm used to the opposite. Not everyone is impressed with family wealth.

MARK SPOKE to his mother about his job search. "Do you think I could work for the company?"

"I don't know. We gave you every advantage. You acted like a common drunk and embarrassed us. You have to earn back your father's trust."

Lew spoke to Pearl about Mark's job search. "It appears that Mark didn't find anything that seemed right to him. I think it would be better for him to have any job, even if temporary, than to be constantly looking. What do you think about my offering him a position?"

"It's a close question. I don't like the idea of interfering in his life. But I agree with you that he has gone long enough without work. He might lose his confidence."

"Any idea what he might do for the company?"

"He's not cut out to be involved in the oil business. He's not tough enough. But maybe finances. You have a company bookkeeper and an outside accounting firm. He can learn from them while working for the company. That's what I suggest."

"Good thinking."

So, Lew offered Mark the job.

MARK WAS surprised by his father's offer. And he had concerns about it.

I'm apprehensive about working for Riggins Oil. I'm afraid that I would be a disappointment to father. But I can't see any other options.

He accepted the job.

He didn't need much compensation. He lived with his parents. When he wasn't at the Club, he took his meals with them. And they paid for everything—clothing, his car, gasoline, and Club charges.

Father will know that the company won't need to make a big investment in me.

He was assigned to the Tulsa office. He took an unused office in the company's suite.

He started by doing menial projects. He ran errands for people in the office. And he delivered things to and from the Yard and to and from people outside the company.

After a while, he was given some responsibilities. He was in charge of necessary office supplies, making sure that they were always on hand. He kept track of the hours of the non-salaried employees. And he sorted the considerable volume of daily mail and got it to the proper recipients.

I like what I'm doing. I'm going to throw myself into it.

People noticed that the office was running more smoothly and efficiently than it had before he came.

Lew, in a rare moment of closeness, showed Mark the family's investment records. He showed him how the investments had grown in value since he had purchased them.

Not taking anything away from father, it appears to me that it's easy to make money, if you have money.

Mark volunteered for additional responsibilities at the company. He had observed that people were having trouble locating files. So, he interviewed the principals, including his father, asking how they set up files and how they stored them. Based on this information, he designed a system that improved retrieval of files throughout the office. And he maintained the files, implementing his system. People appreciated his contribution.

He also asked to assist the bookkeeper, hoping to learn about accounting. With some training, he was able to help with data entry, such as deposits and expenses. But he never achieved an understanding of the company's bookkeeping system or of accounting as a science.

I have communicated that I want to be involved in core management, such as human resources, financial planning, assessing projects, and major decisions. But the principals, including father, have never picked up on the idea. I would like to think this is due to my relatively young age. But I suspect that it's due to a lack of confidence in me, which leaves me feeling inadequate.

He settled into a comfortable routine. He worked Monday through Friday, and recreated Saturday and Sunday.

When not working, he spent time with his parents. And he developed an active social life revolving around the Club. He played golf and tennis and participated in tournaments. And he attended the Club's social activities–dinners, parties, and dances.

He experimented with taking off during the week to play golf or tennis, if the weather was nice. No one seemed to mind if he left for an afternoon. In fact, no one even seemed to notice. So, the experiment ripened into a regular routine.

While I like taking a day off, when I think about it, I feel inadequate. No one seems to notice when I'm gone.

In time, his activities at the company expanded to include "imperial" tasks befitting the son of the major principal. He joined the oilmen's downtown club, of which his father was a member. He attended business meetings. He visited the Yard. And he visited drill sites and sites of operating wells. He met business partners, bankers, lawyers, and company employees. Because of his social skills, he was well received by those he met.

Things were stabilizing for Mark.

I have a good idea of what I'm to do for the company, and golf and tennis are reliable outlets for me.

WHEN THE IMMINENCE of war in Europe became clear, the United States reinstituted conscription. Mark hardly had time to assess the situation before his father had used his influence to obtain a critical-industries deferment for him, based on his working for an oil company.

I understand the concept of a critical-industries draft exemption. And I understand that the oil industry is critical to the war effort.

But I know full well that the exemption doesn't apply to me. I'm not essential to the management of the company. I'm a clerk.

And I 'm not doing essential labor. I'm not a roughneck on a drilling rig. Nor am I a roustabout, maintaining producing wells.

I feel guilty about my deferment. I don't deserve it. But I don't want to be in the Army. I'm afraid of getting killed. So, I'm going to do my best to deal with the guilt.

I wish I had someone to talk to about it. Maybe Edith, at the Club. She was always helpful. But no. I can't bear to mention it to her. I don't want anyone at the Club to think that I'm a coward. Not even Edith. I'll just have to deal with it myself.

―――

FROM TIME TO TIME, Lew worried about Mark.

He seems to have a desire to be unhappy.

Lew and Pearl discussed it, but they decided it was best not to do anything. Best to let Mark make it or not on his own.

―――

HE TRIED NOT to think about the war, but there was too much evidence of it.

The newspapers carried daily lists of Tulsans wounded, missing in action, and killed. Mark couldn't resist reviewing the lists every day.

Are people I know are on the list?

And when he encountered the name of someone he knew, his guilt skyrocketed.

Also, there were constant calls for community involvement in the war effort. They appeared on billboards, in newspapers, and on posters placed everywhere–in schools, churches, storefronts, gas stations, and restaurants. Radio stations also carried them. And people heeded the calls:

They planted "Victory Gardens," to lessen the scarcity of fresh food.

They volunteered to help in hospitals that cared for wounded soldiers.

They collected cans, for use in manufacturing airplanes.

They bought war bonds.

They participated in drives to collect scrap metal.

They saved rubber in all forms—rubber bands, old inner tubes, grommets, kitchen gloves, and rainwear.

Mark did nothing. He didn't participate in any of these activities. He couldn't bring himself to. And this weighed on him when he was reminded almost daily that others were doing their share. He saw front-page pictures in newspapers of children depositing cans and other items with their teachers. And he saw frequent newspaper headlines to the effect of "War Effort Continues!"

Moreover, people lived with the deprivations of rationing—gasoline, tobacco products, foods (sugar, meat, butter, coffee), clothing and shoes, soap, and tires, among other items. But not Mark. If he wanted anything, he just bought it on the black market.

Buying what I want makes me feel even worse. Other people can't afford it. They just have to do without.

I feel completely cut off, isolated, alone. My guilt has led me to be at war with a nation at war.

THEN CAME the Allied invasion of Normandy.

Radio stations and newspapers were full of stories about it. Newspaper coverage was replete with photographs. Allied casualties were enormous. No one talked about anything else for days.

Relatives, friends, fellow workers, and neighbors of those serving in the armed forces in Europe anxiously awaited news about loved ones.

Mark reacted to the news with mixed emotions. *I'm grateful that I've been spared, but guilty that I'm not serving.* All in all, the invasion was a defining event for Mark. It shattered his tenuous hold on his emotions. And his guilt migrated into shame.

THE ADVENT of the war had a subtle effect on the Club's social activities, which was not lost on Mark. Events gradually became fewer and relatively subdued. Reports of casualties suffered by Club members cast a pall over things, a poignant reminder to Mark that he was not serving. Instead of being fun, the activities sent him into a tailspin of guilt, exacerbating his beer drinking, which was the only effective coping mechanism that he knew.

He was ready to find a spouse. But finding a mate was on hold due to an unofficial but effective consensus of the parents of eligible young women. They felt it would be unfair to give an advantage to men who were not serving in the armed forces. This was never articulated to Mark, but he suspected it.

So, Mark was, like most people in the country, experiencing dislocations from the war, even though he was not fighting in it. Rather than be unhappy at the dinners, parties, and dances, he decided to stop attending them. So, he limited his Club activities to golf and tennis.

34

JAKE—RECOVERY

Like Jake and the other men in Jake's squad, the German gunners in the nest were young and inexperienced. And they had been nervous about their initial engagement. Perhaps out of this nervousness, they had not hit any of the other men in the squad.

The other men had witnessed what happened. After the smoke cleared, a spontaneous cheer went up. Relieved and emboldened by what they had seen, they, to a man, jumped up and, braving unseen danger, charged the hill. Rifle against rifle, they easily took the hill, encountering little resistance. It appeared to them that most of the Germans had fled when they saw their machine gun nest destroyed.

Most of the men in the squad spread out along the ridge of the hill, to prevent the Germans from returning. A handful hurried back to see if Jake was still alive and if they could do anything to help him.

Jake was still alive, but unconscious. When he came to, he saw that he was in a field hospital and that a doctor and his aides were stopping the bleeding and administering morphine. He soon went under.

WHEN HE AWOKE, he could see that he was in a hospital room. A nurse was attending to a device that was holding his right leg up in the air.

"Where am I?"

"You're in Mercy Hospital."

"No. I mean what country and what city?"

"Back in the States. Boston. Try to rest. The doctor will be in to see you. You might as well rest until he comes."

Later that day, the doctor awakened him.

"How are you doing?"

"I don't know. I don't feel much of anything."

"That's because you're heavily sedated. Once we cut back on that, you'll feel again."

"What happened? How did I get here? Why is my leg up in the air?"

"You were severely injured in combat. Your right femur, the big bone in your thigh, was badly broken. The field hospital stabilized you, and then the Army flew you here. I operated on you four days ago. Your leg is in traction to prevent you from moving it.

"You're going to be here for quite a while. There'll be a long recuperation. It'll be a few weeks before you can get out of bed. Then you'll need a brace and a cane just to walk. But you'll go through therapy and in the long run you'll be able to walk on your own. You might have a limp when you leave the hospital, but that could go away with time."

Jake thought about what the doctor had told him.

I realize that I have no choice about my course of treatment. I'm going to be patient and do everything they tell me to. But above all, I'm grateful. Grateful to be alive. Grateful to be receiving the care I'm getting. I'm on the way to continuing my life.

THE NEXT DAY, Jake was surprised and thrilled to be awakened by his mother's touch. A younger woman, an Indian whom he did not recognize, was with her.

"Son, this is 'Ronnie Martin'. When the Army notified me about you, she offered to drive me here. I had gotten to know her at a church that I joined just after your father died. And she has been very helpful and generous to me. I don't know what I would have done without her."

Jake could see that Ronnie was embarrassed by Lucy's praise. He tried to comfort her. "It's a pleasure to meet you. Tell me something about yourself."

Before Ronnie could respond, Lucy reached out to him and started crying. "Oh Jake. You look awful. You must have been hurt bad. I prayed day and night for God's protection of you. Are you in pain?"

"No, mom. I'm taking medication for pain. So, I'm doing okay with it."

Jake wanted to take the attention off himself, so he changed the subject. "Tell me about your trip. Was it eventful?"

Ronnie and Lucy jumped into the subject. The trip had been an adventure for them, and they were enjoying the opportunity to recount different aspects of it.

AFTER A BIT, they were interrupted by a knock on the open door. A soldier in dress uniform entered the room, carrying a satchel. "Private Perryman, I have some business to discuss with you." Looking at Lucy and Ronnie, he continued, "would you rather be alone?"

"No, sir. These people came a long way and I want to be with them. I don't want to run them off."

"I will proceed, then." He paused to clear his throat. "I want to first inform you that the Army has granted you an honorable discharge. So, you are immediately entitled to all the benefits that the government extends to veterans. For example, medical treatment and tuition relief." He reached in his satchel and retrieved some papers. He handed a document to Jake. "Here is your certificate of discharge." He then handed Jake a pamphlet. "This booklet explains veterans' benefits."

He then paused. "You were seriously wounded by machine gun fire while advancing up a hill as part of the Allied Expeditionary Force's of Normandy. Because of this, the Army has awarded you the Purple Heart medal." Again, reaching into his satchel, he retrieved the medal and handed it to Jake, who didn't seem to know whether to say something or what to say, when being honored.

Another pause, this one longer than the one before. "You demonstrated extraordinary courage after being wounded. You saved the lives of the rest of the men in your squad, none of whom was even wounded." He choked up for a minute. "Because of your selfless risk, the Army has awarded you a Bronze Star, one of the highest honors it can bestow for valor during combat." Again, going into his satchel, he pulled out the medal and handed it to Jake, who by then was thoroughly embarrassed.

Standing as straight as an arrow, the soldier said, "Private

Perryman," a closing salutation of respect, as he saluted Jake. He then turned on his heel and left the room.

———

NOT HAVING PREVIOUSLY KNOWN what had happened to Jake, or what he had done, Lucy and Ronnie were dumbstruck, filled with surprise and admiration. And Jake didn't know what to make of what had just transpired.

I had no notice of the Army's visit. I have no context for appraising it. I don't know whether it's a standard procedure or something just for me. I've never even heard of the medals.

The three of them were silent.

Lucy broke the silence, quietly weeping. "Oh, Jake. You're a hero! And I didn't even know it. I'm so proud of you, saving the lives of all those men, while you were hurt."

"Aw, mom. Don't make too much of it. Do you remember what dad said when I told the two of you that I had enlisted? 'Do your part, son. Don't leave it up to the other guy.' That's all I did. I did my part. I'm happy to learn that what I did helped the others in the squad. But I just did my part."

Ronnie was beaming at him.

I like the way Ronnie is looking at me, but I'm hard-pressed to know why. Attention usually makes me uncomfortable.

———

THEY EVENTUALLY GOT BACK to small talk.

At one point, Lucy talked about joining a church. "After Sam's death, I got to feeling lonely. He was gone, and you were in the Army. So, I was in the house all by myself. Day after day, alone. It wore on me. I was getting down. I wondered where I could meet people. The only thing that I could think of was a church. So, I visited several of them. Ronnie welcomed me at the door to her church, which had not happened at the others. That meant a lot to me, and the other people I met at her church were also welcoming. So, I decided to return. Eventually, I joined, and I go every Sunday. In addition to being with people, the sermons have helped me to deal with Sam's passing. And there are activities other than the Sunday service. Things like potluck suppers and movie nights. These are opportunities to meet people and have fun. Altogether, it's been—"

A nurse entered the room, interrupting Lucy. As she gave Jake an injection, she politely asked Lucy and Ronnie to leave for the evening. "Jake needs to rest now, and visiting hours are almost over. You can come back tomorrow for more time together."

Jake soon fell asleep. As he did, he was elated. The visit had lifted his spirits.

WHEN JAKE AWOKE the next morning, he still felt the glow of the night before.

I'm grateful that mom and Ronnie took the time and trouble to come see me. I'm looking forward to seeing them again.

The day went well. They talked and talked, mainly about home. Along the way, Jake learned more about Ronnie. She was a bit younger than he. They had gone to the same high school. She remembered his football days. She had a part time job with the county library. She was living with her parents.

In the late afternoon, Lucy and Ronnie could see that he was tiring. They said they should leave soon. They explained that they needed to start back home the next day, and they wanted to get an early start. They assured him that he would be home before he knew it and that they would be looking forward to his return. Lucy kissed his cheek and they moved toward the door. Just before leaving the room, Ronnie turned around, smiled, and waved goodbye.

After they left, Jake felt something that he had never felt before. Empty.

Something 's amiss. It must have to do with mom. Because of all that's happened, it seems like it's been a long time since I've seen her It was a short visit; and there's no prospect of seeing her again for months.

———

AFTER A COUPLE OF DAYS, the excitement of the visit wore off, and Jake focused on his recovery.

I want something to help me pass the time as I heal and undergo therapy. I also want to improve myself.

So, the hospital helped him sign up for a correspondence course in applied physics, given by a local university.

He worked on the course and on his therapy the way he had always done things. He paid attention and gave them his best.

The results were salutary. He got a top grade in the course and he surprised his doctor and the staff by being ready for a discharge sooner than expected.

The Army sent him a fresh dress uniform, a duffle bag, and necessities for a week; it booked passage for him on a train to Oklahoma; and it told him that it had notified his mother of his arrival date and time.

The hospital gave him a cane as a precaution, although he was walking quite well without one.

I'm ready and anxious to return home.

35

RONNIE

Reporters and photographers greeted Jake as he got off the train in the Tulsa depot. He was uncomfortable, but he managed to say that he was very happy to be back home. They pressed him for details about what had happened on the battlefield and how he had felt about it.

He was starting to get angry, but then he spotted Ronnie on the platform. He was expecting Lucy, and his heart skipped a beat when he saw Ronnie. He smiled and waved to her and she smiled and returned his wave. He temporarily froze. The memory of her leaving his hospital room came rushing back.

I now understand my empty feeling after Ronnie and mom left my room to return home. It had nothing to do with mom. It was due to Ronnie's departure. I now realize that I had been smitten with her.

"Sorry guys. I have to go. My ride's here."

He cut through the reporters and rushed to her, and she moved forward to meet him. He looked deep into her eyes and wanted to touch her, but he couldn't bring himself to do it. She looked him over, head to foot. "You're handsome in your uniform."

As they walked to her car, he quipped that she had rescued him. She laughed. "You didn't need rescuing. You've more than proved that you can handle yourself." She then explained that his mother had wanted to be there, but that she was not feeling well.

I'm sorry about mom, but I'm excited to have some time alone with Ronnie.

"It's awfully nice of you to pick me up, even if you didn't rescue me."

"When your mother called, I was happy to do it. It would give us a chance to get to know each other a little better."

They stowed his duffle and cane in the trunk and got into the car.

Jake was nervous.

I've always been shy with women and, as a result, haven't been successful in winning their attention. Now I have Ronnie's attention, something I would not have dared to wish for. I can't believe my good fortune. But what to say?

His heart was racing, he was perspiring, and he was hyperventilating.

"How about telling me about your church? Are you pretty active in it? We never went to a church, so I don't know much about them."

She explained that she had not grown up in a church but got interested in one after high school. She liked the community, and the spirituality had come to mean a lot to her. She taught a Sunday school class for a term. Now she was a greeter. She met people as they entered the sanctuary on Sunday.

"Maybe you could try coming to church with your mother. See how you like it."

I can't believe what's happening. She's suggesting something that'll be a way for us to see more of each other—exactly what I want. *I can't know whether that's the way she sees it, but it might be. I'm in another world. The future looks bright.*

To fill an awkward silence, she asked some questions about him, and they were soon at his house.

"Thanks again for the ride. I guess I'll see you Sunday."

JAKE WENT to church with his mother that Sunday. He was apprehensive about whether Ronnie would have time for him, given that she needed to attend to other people. But it turned out to be a baseless concern. She greeted him warmly at the door. And she sought him out during the coffee hour after the service.

They had a nice conversation.

So, he formed a habit of going to church every Sunday, and he made it his business to visit with Ronnie when he was there. And he pushed his luck by participating in some of the church activities that she liked.

In addition to paying attention to Ronnie, and closely related to that pursuit, he looked for a job.

I want something solid—full time, and not seasonal or temporary—so I can get started settling down and saving money.

That's what he got. A job with the Parks and Recreation Department of the city of Tulsa.

ABOUT TWO MONTHS after he got home from Boston, he got an invitation to an American Legion dinner dance. He summoned up his courage and asked Ronnie to go with him. To his relief, she accepted.

When he picked her up, she answered his knock on the door. She was dressed appropriately–a black slip dress and heels. And her hair was done up. He stared at her, speechless. She asked, "Is something wrong?" He replied, almost stuttering, "No. It's just that you're really pretty tonight. I mean you're always pretty, but tonight is different."

"Thank you. But there's no mystery. I don't go out often, so I didn't want to miss an opportunity to dress up. But I appreciate your noticing."

They made a handsome couple, with her dressed for the occasion and him in his dress uniform, with medals. More than once, Jake saw other guys ogling Ronnie.

I'm proud that guys think she's attractive. But I'm also jealous and possessive. I want her all to myself. So, I have to watch myself. I don't have a claim on her attentions. I don't want to say or do something inappropriate. But I do want to make a move at an early moment to solidify our relationship.

He really enjoyed the dancing. The slow numbers permitted physical contact that was new to them. When her cheek occasionally happened to touch his, he could not help but notice.

When they got back to her parents' house, he walked her to the small porch adjacent to the front door. She turned to him. "Thanks, Jake. It was a great evening." He started to reciprocate, "I enjoyed it too. I'm glad–" But as he was speaking, she inched forward and tilted her head back, movements so subtle that they were almost imperceptible. Ordinarily, he wouldn't have noticed, but due to heightened awareness, he sensed the movements and felt sure of himself. He embraced her, and their lips met.

He had kissed girls before, but those kisses meant nothing compared with the kiss that night. That kiss got them over a hurdle, due to his timidity, that they had to clear for their relationship to progress.

And progress it did. They started going out for dinner and, occasionally, to a movie. They expanded their knowledge of one another by getting away from the structured personal relations in the context of church activities.

Jake's expression, both verbally and physically, of his feelings toward Ronnie was often awkward. But her woman's intuition smoothed things out. And it was only a matter of time until they married.

36

A FAMILY

They could not afford housing. So, at Lucy's invitation, they established their marital home in Lucy's house. Lucy liked the arrangement, but it inevitably confined privacy to the bedrooms. This sometimes produced stress for Jake and Ronnie, but they were patient in working through it.

One evening, Jake was noticeably preoccupied by something when he got home from work. Neither he nor Ronnie said anything about it until they were getting ready for bed.

"What's on your mind, Jake?"

"I can't do anything about it. But I got passed over for a promotion today. They gave it to a white guy with less longevity than mine."

"Did anyone explain why?"

"No. But I know the reason. I'm an Indian and he isn't."

"I hate to hear that."

"I know. But it's just the way things are. It's no different now than it was when our parents were young. We're still second-class citizens. We're different. We're not entitled to the things that whites take for granted. Someone told me a couple of weeks ago that some bars in Tulsa still have signs that Indians are limited to two drinks."

"I agree with you that there's nothing to be done about it. And I'm sorry you didn't get the promotion, but we'll just have to live with it."

This development disappointed them, but they took it in stride. And their day-to-day marital patterns continued to evolve.

WITHIN A YEAR, Ronnie became pregnant. Jake was excited. He observed, without comment, changes in her body and behavior.

He was solicitous of her to the point of her amusement.

She gave birth to a girl, "Lucinda."

Lucinda had a crib in her parents' bedroom. The room was crowded, but no other place in the house would work. Aside from two bedrooms and a bathroom, there was a largish kitchen with a table for meals and there was a living room where Ronnie, Jake, and Lucy spent most of their time.

Jake told Ronnie, almost apologetically, "The arrangement is temporary. I plan to borrow some money and add a third bedroom. It would be small, and I could do a lot of the labor, holding down the cost."

THE PLAN TURNED out to be unnecessary. When Lucinda was just three months old, Lucy died in her sleep from a massive stroke.

A modest church funeral would take almost all of Jake and Ronnie's savings. But they both wanted to have it. The alternative was a county burial for the indigent.

As Ronnie was getting Lucy's room ready for the funeral home to remove the body, she discovered cash in a drawer. Lucy, ever frugal, had managed to save some of Jake's Army pay, which it had sent to her every month. And her skepticism about banks had been vindicated. She had not lost her savings to a bank failure.

"Look, Jake. This was in a drawer in your mother's room. I counted it, and it would cover the funeral. I suggest we use it for that."

"I agree. That seems fitting." Jake was moved. "I'm grateful that you feel that way."

She was buried in the cemetery of the tribal community. Her grave was next to Sam's.

———

IN TIME, Jake asked Ronnie to clear and clean his mother's bedroom. He told her that he was not up to it. He was still grieving her passing and disposing of her things was frightening to him.

It seems like a taboo undertaking, but I know it's not.

Ronnie took care of it. And she went further.

She thought, *I know Jake's feelings. I'm going to change the room so that he no longer thinks of it as his mother's.*

She visited a second-hand store and worked out an exchange. She traded all the furniture in Lucy's room for used children's furniture. And she painted the room in colors that matched the mood of the furniture. She ended up with a cheerful room for Lucinda that would be appropriate for her for years.

Jake was enthusiastic about the steps that Ronnie had taken.

"I'm pleased with the results. I feel that the house has been renewed and is comfortable for us."

And Ronnie's changes brought him closure with his mother.

I continue to miss her, but I no longer associate the house with her.

It's now our house.

THE FAMILY WAS NOW FULLY FORMED, and day-to-day routines set in. Jake went to work. Ronnie stayed home. She did the housework and, until Lucinda started school, cared for her.

On weekends, Jake took care of the lawn and the house, making repairs and painting it inside and out.

He was a good father.

When Lucinda was very young, he made up stories for her. He told them over and over at her bedtime. She loved them. Three of her favorites were "The Owl and the Hootenanny ", "Alley Opp Opp Spilled the Soup, and the Beans went 'Plop' on the Floor", and "Freddy, the Flying Wombat, the only Ornithologist in Town". But her all-time favorite was "Hippo Loto Latamus, the High Wire Hippo, Star of the Circus." Sometimes she got so excited by the stories that Ronnie had to come in and put an end to it, so she could sleep.

As she got a little older, he spent time playing with her, attending birthday parties and other special events, and going on outings. Sometimes, he and Lucinda would sit in front of the radio and listen to Saturday programs for children. Every other weekend, he and Ronnie took her to visit her maternal grandparents.

37

HOME IMPROVEMENTS

J ake wanted a substantial improvement to their house. *I feel bad that I wasn't able to provide a house for us when we had just gotten married. I'll make it up to her now by improving the house that we now own.*

"Ronnie, I'd like us to have running hot and cold water in the house. It would make life more enjoyable and much easier."

"I agree. But can we afford it? We've closely watched our expenses, but our savings are quite small."

"I think we should borrow for it. I can do part of the labor, to hold costs down.

Jake checked out a book on plumbing from a library in Tulsa. And he sought advice when he needed it.

Over the next month, Jake worked on the project evenings and weekends. He took measurements, did computations, and prepared a floor plan.

We'll need a water heater and a septic tank. Maybe a new bathtub. *We have a good water well nearby.*

He selected a contractor and met with him. They reached an

agreement on how much Jake would do. And the contractor estimated the cost of the entire project, including his fee.

Jake and Ronnie went to Sam's bank and arranged a loan. The bank set up a special account for the money.

Work began immediately.

Because of the amount of work that Jake had agreed to do, he had to occasionally take off from work.

This bothers me, because we're now in debt. But it's necessary.

They finished in a little more than three weeks.

Jake wanted a ceremony, so he got Ronnie to watch while the contractor turned on the electric pump for the water well. Jake was a bit nervous, but everything worked perfectly.

It didn't take long for Ronnie to become really excited about all the nice things that running water brought to their daily lives. "I can't believe it. How nice it is to be able to go to the bathroom indoors! And I can't believe how much easier things are by not having to pump, heat, and haul water, especially for bathing and for washing clothes and dishes."

Hugging Jake, "You're a genius, Jake. I'm forever grateful."

Jake couldn't help but compare this project with helping his father with the farm.

I was the prime mover for both projects. Of course, there were big differences. The largest was all the uncontrollable factors involved with the farm—the weather, the Depression, the size of the debt. While the running-water project was less complex, it involved a lot of work by me and was a complete success. I conceived of it, did the research, drew the plans, and did a lot of the labor. And it all works well. I can justifiably be proud.

THE CITY CONTINUED to pass Jake up for promotions, giving the positions instead to white men with less seniority. The absence of promotions limited his pay to annual, within-grade raises, which were modest and were the same for everybody.

Jake and Ronnie were unable to save and were chronically short of money. Loan payments, house upkeep, three to feed and clothe, and occasional help for her parents took all they had.

Occasionally, Jake attempted to find a better, higher-paying job. But race thwarted him. He was a veteran, and a decorated one at that, but he was still an Indian.

During a period when many Americans were buying new cars and appliances, that was unthinkable for Jake and Ronnie. They often had to make do or even do without. They were part of a new class—the working poor. Yet they were a happy family.

JAKE HAD a change of fortune at work. Another worker was seriously injured on the job. The City had to rearrange things to cover for him while he was out. Jake volunteered to do more than the work of one person. And he made useful suggestions on how to fill the gap. His boss was pleased with both his attitude and his contributions toward covering the loss. So, one Friday evening his boss handed him an envelope containing an unexpected bonus.

Ronnie spoke first. "We can use it to make an extra payment on the loan."

"I'd rather use it to further modernize our home. I'd like to replace our ancient wood-burning stove. That's the other project I've had in mind. I see the bonus as found money. I'd like us to use it for something nice that we otherwise couldn't buy."

"That's an interesting approach."

"Imagine not having to split wood anymore. Wouldn't that be

nice? And it would bring our house up to date. We have modern plumbing. But we still have an outdated range."

"Do you think we could do it for the amount of the bonus?"

"Let's find out."

Jake and Ronnie went to a store that sold ranges. Due to the age of their house, it had minimal electric service, mostly for light bulbs. So, the manager advised them to get a gas range. They found one that they liked. The cost of the range, and an outdoor propane tank, both with installation, was less than the bonus.

38

LUCINDA GROWS

As Lucinda got older, life became more complex and busier for Jake and Ronnie. Lucinda enlarged her activities and joined groups. She signed up for elementary school softball. And she volunteered to be in a school play put on by second graders. And there were social events of all types—church picnics, school outings, and spur-of-the-moment plans to join other children to go to a Saturday afternoon movie, to go to an ice cream parlor, or to go to a local petting zoo.

Ronnie and Jake divided the driving duties between them. During the day on weekdays, Ronnie drove. During evenings and on weekends, Jake drove. While Jake enjoyed his part in these activities, they resulted in some lost sleep and crowded weekends.

Of all Lucinda's activities, Jake was most interested in the soft ball. He helped organize the practices and games. And it wasn't long before he was coaching the team. Ronnie was often in the bleachers.

JAKE VOLUNTEERED to teach a Sunday School class. He was an immediate hit. From a list of suggested courses, he chose to teach about King Tut.

My goal is to both teach and entertain. I want the children to use their imaginations to see what it would have been like to live in a strange and far-off country 3,000 years ago.

He wrote and directed a play about King Tut's life. Each child had a part and devised a costume for the part. They performed for the whole congregation one Sunday morning, which was a huge success. Ronnie was in the audience.

I'm so proud of him. He is great with children. And he is able to do anything he puts his mind to.

It was not unusual for a child to stay back after the class ended. After some timidity, the child would tell Jake about a problem he or she was having. Jake sat down and motioned for the child to sit. By listening patiently, and by asking the right question at the right time, Jake was able to help the child cope with the problem.

39
TRADEGY

After work one night, Jake was driving westbound on the Interstate.

An eighteen-wheeler was eastbound on the same road. The driver was groggy, but he had decided to keep going until he got to a truck stop that was just east of Tulsa. It was a fateful decision. He fell asleep as he neared an overpass. The truck hit the bridge abutment so hard that the impact broke a fuel line. And sparks from metal scrapping against cement caused a fire under the hood.

Jake got there just after the accident. He realized the danger.

The engine compartment's on fire; the driver's not getting out; the fuel tanks could blow at any moment.

I can't drive on, as if I hadn't seen it.

He stopped, got out of his truck, ran across the Interstate, vaulting over the guard rails on the median, and mounted the step to the cab's door. He tried to open it, but the impact had jammed it. While he was struggling with the door, the fuel tanks exploded, instantly killing both the driver and Jake.

THE ARMY CONDUCTED a funeral in a military cemetery, where Jake was buried.

He was 43 years old.

Ronnie was grief stricken. For two weeks, she thought about him continuously.

We were so much in love. Our initial feelings toward each other never grew stale. I'll miss him terribly.

He was a great father. How can I finish raising Lucinda without him?

He overcame his poverty. He was born poor, lived poor, and died poor. But he lived well.

She cried inconsolably for those two weeks. After that, for Lucinda's sake, she worked hard not to prolong her grief.

40

LEW—JOURNEY'S END

Lew never forgot how he had gotten started. A couple of times a year, he visited the Perryman-1 well, in the 40 acres he had gotten from Sam. It was almost like visiting a gravesite. He would sit for a while and think about how things had worked out.

I wonder what things would have been like if this first well had not been good. I know that I would not have tried a second time. We didn't have the money to do it. So, this well is the source of everything I have accumulated.

He always walked away feeling grateful.

The dream that pulled me from Des Moines to Indian Territory has been fulfilled. I'm satisfied with my choice.

LEW DIED SUDDENLY and unexpectedly of a heart attack

His death devastated Pearl. They had been very close. They were more than husband and wife. They were extraordinarily successful business partners.

She did not have a family when they met. He had provided her with one–his family in Iowa.

It took her years to get through her grief. And she never again participated in social affairs. She became a recluse, except for her expanded participation in the company.

THE FUNERAL WAS PACKED, which was not surprising, given the number of people Lew had encountered through his work and other activities.

Pearl had asked Dan Harrington, the company's geologist, to speak at the service. He spoke about Lew's decisiveness, his willingness to take risks, and his ability to delegate. He wound up his comments:

"He was well known. He was well liked. He was well respected.

He amassed a fortune.

He was a legend in his own time."

41

MARK—WORLD WAR II

Then his father died unexpectedly. His first reaction was shock. His second, coming soon after the first, was panic. He had relied on his father for everything—college tuition, housing, country club privileges, and his job.

The latter bothered him most.

Will the other owners fire me? If not, will they look to me to run the company? I don't know much about it. How can I run it?

His drinking became less sporadic and more frequent. When he got home from a worrisome day, he would sneak a drink. He couldn't afford for his mother to see it. But it was a large house, with plenty of places to hide a bottle.

After an appropriate delay, the three remaining board members counseled with Pearl, who had inherited Lew's assets, including his company stock. They hired an outsider to replace Lew as CEO. And they instructed him that Mark's role at the company was to remain unchanged, except that he was to be named Chairman of the Board, with the understanding that it would be only a title. He would not be in charge of the real officers.

Pearl explained all this to Mark.

"Son, I've been meeting with the other owners of the company, which is going to happen regularly from now on. I'll join the board and will oversee the company's operations. We're going to hire an outside person with a lot of business experience to replace your father. To assure our employees and the public that the company will remain unchanged, you will be named Chairman of the Board. You will be seen as carrying on for your father. This means that you'll need to attend more meetings and increase the time that you spend relating to employees. Does this seem all right to you?"

"It's a little frightening. I wonder if I'm ready to be Chairman of the Board. I mean, being in charge of other people."

"Don't worry about that. The Chairman of the Board is usually in charge of the other officers. But in this case, the other officers will operate independently of you."

"But what about meetings? I doubt that I can handle meetings the way father did. He knew all about the company. And he could make decisions during a meeting, because he understood what the company needed from the meeting."

"We've discussed that. You'll have whatever help you need. Another officer, or

the company attorney, or its accountant, will be with you at meetings. You won't be asked to make major decisions. Does that put your mind at rest?"

"Yes. It does."

"All in all, I think you'll be comfortable with the arrangement. And you'll be doing your part by showing that there's continuity in the company's management."

"Well, I'll certainly do my best. I hope people will let me know if I need to improve."

"Of course. We understand that you'll be doing things for the first time. We're all here to guide you and make suggestions where appropriate."

"One thing I'm wondering is whether I can learn more and become more than a figurehead."

A pause. "Let's see what develops. Okay?"

I feel inadequate. Even my own mother doesn't think I can do that much for the company.

"Sure. I'll appreciate whatever opportunities come along."

"And one other thing. I ask you to be careful about your drinking. You have more responsibilities now than before."

"I take a drink every once in a while, mother, but it hasn't been a problem."

"I didn't say it has been. I just mention it because I want it to stay that way."

So, Mark began his new role.

MOTHER WAS RIGHT. My new position isn't too challenging. And I like the additional attention I'm getting in the business.

He got caught up in his new duties. For the first time, he became enthusiastic about his work. For a while, he stopped feeling inadequate about his job.

HE CONTINUED the drinking patterns that he had fallen into at college, except that he gradually changed from beer to Bourbon. Bourbon, having higher alcohol content than beer, caused earlier intoxication. But his drinking remained sporadic, mainly at social events.

There was a logistical problem with drinking bourbon. Oklahoma was a dry state. There were no liquor stores, and public bars could serve only beer. So, Mark had to find a bootlegger, which was not difficult, because so many people relied on them.

However, because he was living with his parents, he couldn't ask the bootlegger to deliver the bourbon to his home, which was the normal practice. Nor could he have it delivered to the office. So, Mark arranged to meet the bootlegger at an out-of-the-way neutral place, a place where no one would likely notice what was happening.

The law allowed him to check a bottle at the Club. To avoid being too obvious, he ordered just one drink a night. He supplemented his checked bottle with a flask. When he was sure no one was looking, he refilled his glass. Then he asked the waiter for more ice.

HE CONTINUED the drinking patterns that he had fallen into at college. His drinking remained sporadic, mainly at social events.

There was a logistical problem with drinking bourbon. Oklahoma was a dry state. There were no liquor stores, and public bars could serve only beer. So, Mark had to find a bootlegger, which was not difficult, because so many people relied on them.

However, because he was living with his parents, he couldn't ask the bootlegger to deliver the bourbon to his home, which was the normal practice. Nor could he have it delivered to the office. So, Mark arranged to meet the bootlegger at an out-of-the-way neutral place, a place where no one would likely notice what was happening.

The law allowed him to check a bottle at the Club. To avoid being too obvious, he ordered just one drink a night. He supplemented his checked bottle with a flask. When he was sure no one was looking, he refilled his glass. Then he asked the waiter for more ice.

42

MARK—POST WAR

Within a year of Lew's death, the war ended. Mark was more than ready for it. He had reached a point where he wouldn't talk about the war. It brought on more shame. He couldn't even bring himself to attend the victory celebration in downtown Tulsa.

I don't deserve to join in.

But he found that the end of the war was not the end of his shame about it.

When I see someone who returned maimed, or when I talk to people with loved ones who are missing or dead, the shame returns.

Nevertheless, Mark was pleased that after the war had ended, the Club returned to normal very quickly. Social activities increased markedly and were fun again.

A month after the end of the war, the Club scheduled a dinner to honor returning vets. After dinner was done, and tables cleared, the Club chairman walked to a podium and began the program.

"Welcome. And thank you for attending this celebratory

In Indian Territory 177

dinner. We can all be pleased with the large turnout, to honor these young men who valiantly served our country in the Great War. Borrowing from Lincoln's Gettysburg Address, 'It is altogether fitting and proper that we should do this'"

"I ask all the honorees to stand up. You will notice that each of them is wearing the dress uniform of the branch in which he served."

General applause.

He began the program by reading a list of those known to have lost their lives. Then he called for a moment of silence.

Next, he read a list of the missing in action. And again, he called for a moment of silence.

Mark was fidgeting in his chair.

I shouldn't have accepted the dinner invitation. People, especially women, are staring at me disapprovingly, as if to say, "You shouldn't be at this event. You should have been fighting, just like the others." And I have to agree with what they're thinking. The program is already dragging me back into shame.

Then the chair called on each vet and spoke about each.

"David Lucas, would you please stand?"

General applause.

"David volunteered for the Army in 1942. He was assigned to the 16th Infantry Regiment, US 1st Infantry Division. On June 6, 1944, he participated in the Allied Invasion of Normandy. He went on to be part of the fighting force that liberated eastern Europe. He was wounded toward the end of the fighting. He received an honorable discharge and was awarded the Purple Heart medal. David, take a bow."

General applause.

Mark continued to be uncomfortable.

Why did I accept the invitation?

I thought I had to attend; I am now the figurehead of Riggins Oil. But I can see that that was not the real reason. The real reason was that I wanted to prove to myself that I had gotten past my shame about my draft exemption. But it didn't work.

I need a drink.

Mark ordered the first of what would become a number of bourbon highballs.

The program continued for more than an hour. Then the Club chairman wound it up.

"Again, thanks for attending. The program has been very meaningful for me. I hope it's been the same for you. Feel free to mingle, now that we're done."

Pearl's driver arrived at the regular place, at the time Mark had suggested. But Mark wasn't there.

After waiting for a while, and seeing that people were leaving, the driver went inside to look for Mark. But he couldn't see him. So, he asked the Club manager for help. They searched everywhere, including the two men's bathrooms. Then they went outside.

They found him on the veranda, passed out on a chaise.

THERE WAS another change at the Club, after the end of the war–the hold on finding a mate disappeared.

Young women in the Club vied with each other to sit next to Mark at dinners. He charmed them, because of his social skills. But nothing clicked.

Everything changed the night that Nora sat next to him, which her brother had arranged with the Club manager. She was beautiful, and she was charming. She listened well. He was smitten.

I can't get her out of my mind. I thought about her last night, after the party. And I've been thinking about her today. I want to see her again, soon.

So, he called her and arranged a dinner date for that night. It reinforced his initial reaction. They talked and talked. He was charmed.

I want more contact with her.

The Club had a dance two days later. Mark and Nora agreed to attend as a couple.

Holding her while they danced was another high for him. Soon they were dancing close to each other. The change was not very perceptible, but it was there and they both knew it.

As they left the floor after the last dance, he tentatively reached for her hand and loosely interlaced his fingers with hers. Her response was anything but tentative. She tightened the grip.

They kissed that night. It was gentle, but a beginning. Soon, their farewells turned into passionate embraces.

Something unfamiliar is stirring in me. I think it's more than a passing fancy. But I'm not sure.

So, he spoke to his mother. She helped him realize that he was in love with Nora and wanted to marry her.

"Thanks for your help, Mother." Hugging her, "Now I feel certain about what I want to do."

Pearl thought, *He's so naive.*

Things moved rapidly after that.

It was a big, formal wedding. The local newspapers billed it as the social event of the season.

After their honeymoon, they moved into a grand house that Nora had picked out and decorated. It was a wedding gift from Pearl.

In the excitement of his courtship and marriage, he was able to moderate his drinking. He didn't drink excessively the whole time.

For the first time since high school, he felt comfortable, even elated. He didn't feel like he was on the outside, looking in. Most people got married. And now he was married. Sometimes he had to remind himself that it had really happened.

———

FIVE MONTHS AFTER THE HONEYMOON, Nora greeted Mark as he came in from work.

"I'm pregnant."

Mark was caught off guard

I think I feel good about the news, but I'm not sure. How will it change things?

Then he was thrilled in a way he had never been before.

He fumbled for words. "That's great!" A pause. A long embrace. "Do you feel okay?"

"I've been a little nauseated the last few mornings. The doctor says that's normal and that it will go away in a couple of months."

"We'll need to get a room ready. Repaint it. Get some furniture. Would you like to do these together? I'd love to."

"I would too. It'd be a good way to keep reminding us that it's really going to happen—I'm going to have a baby."

It was a girl. They named her Karen.

Mark was excited about the baby. He couldn't get enough of holding her and even of changing her diaper. When she became restless in the middle of the night, he enjoyed getting up and rocking her.

And he waited on Nora hand and foot.

"I'm afraid you're spoiling me."

"Not a concern. You'll feel better and stronger in a few weeks. Then I'll wean myself from taking care of you."

During the excitement of having a new baby, Mark had a drink or two from time to time, especially after Nora got to sleep. But heavy drinking didn't happen.

Inevitably, however, the excitement faded. And when it did, what was left was Mark. Nothing had changed.

I'm still haunted by feelings of inadequacy and shame.

Nora noticed the shame, and she decided to say something about it.

"When people talk about the war, you become silent and you shy away. You often leave the room. You've obviously got strong feelings about it. Have you thought about getting help to deal with those feelings—a clergyman or even a psychiatrist?"

Gruffly, "No. I don't need any help. The war's over. Over time, my feelings about it will go away. I'll get them behind me. I'll quit thinking about the whole thing without help."

But he never did. And his shame—along with his continuing feelings of inadequacy—contributed to a rapid return to heavier drinking.

Nora decided to say something about it.

"It seems to me that you have really increased your drinking lately. Do you want to talk about it?

Mark said nothing.

Nora sighed.

Pearl also noticed the change. She confronted Mark.

"Don't you think you're drinking too much?"

"Oh, I've gotten drunk a couple of times. But nothing came of it."

"From what I've heard, it's been more than a couple of times."

"No, mother! It hasn't. I ought to know. So, you can stop thinking about it. I'm all right."

But when Pearl and Nora got together, ostensibly for grandmother to bond with granddaughter, the talk quickly turned to Mark's drinking.

Pearl asked, "Do you think Mark drinks too much?"

"Yes, I do. And I glad you asked. I was afraid to ask you."

"Why is that?"

"I'm his wife. I'm with him daily. It seems like I should be able to do something about it. But I don't know what to do. And I've been afraid that you'll judge me harshly for that."

"Not at all, darling. Mark's hard-headed. If he wants to drink, he'll drink. Your job as his wife is to cope with him. Have you tried talking to him about it?"

"Yes. Twice. Both times he said nothing, just sat there without saying anything. And just telling you this makes me feel ashamed."

"Nonsense, Nora! Shame isn't appropriate. It's a difficult situation. And I'm sure you're doing your best to deal with it."

But in fact, Nora couldn't do anything with Mark.

I don't understand his relationship with alcohol. I've never before encountered alcoholism and I don't know how it works.

MARK WAS BECOMING bored with work.

Just being nominally involved is getting old. All the business meetings seem the same.

And I'm tired of going to the lunch club for oil men. I don't have anything to say to the people there. My knowledge of the oil business is superficial. They're deeply involved in all aspects of it.

Visiting the Yard and the well sites isn't bad, but I think the employees are tiring of it. I get the feeling that they view it as interrupting what they're doing and wasting their time. They've come to understand that a Riggins is still involved. Why labor the point?

He began drinking at dinner every evening.

He always drank when he and Nora went out. Sometimes he got really drunk, embarrassing both Nora and the people with them. When this happened, she insisted on driving them home. Mark got angry and argued with her. But he always gave in.

Gradually, he and Nora limited their social activities to the Club, where he didn't stand out so much. They kept up with it, because they both felt that they needed to keep up a good front. She always drove.

He had become obsessed about not running out of bourbon. He had a second bottle in his office, in case he finished the first. He hid bottles around their house. And he kept several bottles in the trunk of his car, wrapped so they would not rattle.

WORK WAS GETTING INCREASINGLY UNPLEASANT.

Other people are doing things. I'm just acting.

He became negative about what he was doing.

He started drinking at lunch. If he didn't go out to eat, he had one or two drinks from the bottle that he kept in his desk.

It wasn't long until he had a drink or two every afternoon. And he started nodding off and napping every afternoon.

People could smell alcohol on his breath after lunch. They started working around him. They no longer asked him to participate in afternoon meetings, just morning meetings.

PEARL STEPPED IN, which was unusual.

"I want you to stop visiting the Yard and well sites."

"Why is that, Mother?"

"Because of your drinking. People can smell alcohol on your breath. That could damage the company. Employees might fear for their future, if they thought the company was being run by someone who drinks too much."

"I'll do as you say. But I don't think it's a problem."

"I do. And so do the other directors. That makes it a problem."

"OK. OK. Fine with me."

"I told you that you were drinking too much, but you didn't believe me. Why don't you do something about it? Maybe see a psychiatrist."

"I don't have a problem. I just take a drink every once in a while. I'm not sitting under a bridge drinking wine from a bottle in a paper bag."

HE STARTED DRINKING EVERY NIGHT. After Nora and Karen had gone to bed, and he had kissed them goodnight, he stayed up drinking, alone.

He lost a diversion from the office when he stopped playing golf and tennis. No one wanted to play with him because of his daytime drinking.

Mark and others at the office didn't know it, but Pearl had developed congestive heart failure. She was a very private person and didn't discuss her physical condition with others. And it was hard to pick up on it because she had gotten to the point of rarely leaving home. Except for domestic help, with whom she had always been aloof, for the most part she sat at home alone.

When Mark visited, he didn't observe anything out of the ordinary. She had gained weight over the years and was always a little out of breath.

She was disheartened by Mark and had lost interest in him. They were not close any more. And she had lost contact with the company.

Her death was a surprise to both Mark and the people in the

company.

Her lawyer explained to Mark that she had left everything to him. The Tulsa newspaper reported that he was one of the wealthiest men in Oklahoma.

The inheritance means nothing to me. It won't change me or my life. I know that it's all due to father; none of it is due to me. Father had been supporting me all my life and still is. I have everything I need or want. What difference does owning the family assets make?

HE WAS PAYING LESS and less attention to Nora and Karen. He was often sullen and nonresponsive at dinner. He was so enmeshed with himself that he couldn't think of anything to say. Nora tried without success to draw him out.

"How are you, Mark?"

"Oh, I'm fine."

"Did you have a good day?"

Shrugging his shoulders, "The usual. Nothing new."

Mark attended Karen's fourth, fifth, and sixth birthday parties.

I like the parties. I enjoy teasing Karen's friends. And it's obvious that they like it. It gives them a chance for some attention, instead of having all the attention focused on Karen.

He made it to Karen's seventh birthday party. But he didn't participate. He was distant, compared to the way he had always been at her parties.

AS HIS NIGHT drinking got heavier, his hands started shaking in the morning. He stopped eating breakfast and took a drink or two, before leaving for the office, to stop the shaking. He kept a bottle hidden in the den for this purpose, and he was careful to make sure he was alone when he poured the drinks.

People noticed alcohol on his breath when he arrived in the morning. They reacted by cutting him out of everything. They didn't include him in meetings. They diverted his mail to other people. They stopped giving him reports and proposals.

MARK WAS INTOXICATED at Karen's eighth birthday party. He passed out, with his head on the table. Nora woke him, with difficulty, and helped him get onto their bed, fully dressed.

He was doing nothing at work. He came in early. His desk had nothing on it except family photographs and paperweights. He sat alone in his office all day, looking out the window. He drank. He napped in the afternoon. Before going home, he checked to see if he needed to dispose of an empty bottle and replace it with a full one.

He missed Karen's ninth birthday party. He was preoccupied with discreetly disposing of empty bottles that had accumulated in his car trunk. He felt that it wouldn't do for people to see an empty bottle at the office or in their trash at home.

Because of his stock ownership, Mark could have made changes in the company, to give him a larger role to play. But he didn't even think about it. He had lost interest in the company, just as he had lost interest in golf and tennis, and even in Nora and Karen.

OVER A PERIOD OF YEARS, Mark's drinking had escalated. It was not a regular progression, but a steady one.

Beer
Bourbon
When going out after dinner, sometimes and then every night
Before dinner, sometimes and then every night
Plus, at noon
Plus, in the morning
Plus, in the afternoon
Ultimately, he was drinking all day long. His only interest was alcohol. His life had become centered on it. He was comfortable only when he had alcohol in his system.
Comfort at a cost. A high cost.

HIS THOUGHTS WERE TURNING deep and morose.
I didn't try out for football in high school. I was afraid of getting hurt.
I didn't apply myself to my education. I don't have a college degree.
I gladly accepted a draft exemption, even though I knew I didn't deserve it.
I gave up my dignity by accepting a figurehead role at the company. It was easy to do what paid advisors told me to do.
I'm not doing anything at work. If I go to the office, I just sit there, doing nothing.
I've been chasing my father's stature my whole life. I've never even gotten close. I have accomplished nothing.
My friends have real jobs, make decisions, and accomplish things.
I've lost interest in everything.
I'm living a pointless life.

SPRING CAME EARLY THAT YEAR. Clear skies; mild temperatures. The Club's social committee set the date for the annual Spring Gala. Plans centered on a festive dinner, accompanied by a string quartet.

Mark was more talkative than usual that night. Nora observed this and was pleased. He seemed to be relieved of the demons that had been plaguing him.

About thirty minutes before the party was scheduled to end, he stood up and excused himself. "I need a breath of fresh air."

He headed to the door that opened onto the veranda at the back of the main clubhouse. On his way, he shook several offered hands and exchanged pleasantries.

As the party was winding down, Nora looked for Mark, but couldn't find him. She asked the couple they had come with for help. They searched other rooms in the clubhouse and the men's rooms. They figured that he was outdoors and had lost track of the time. So, they asked the Club manager for help. He called on the waiters and kitchen staff to fan out and search for him.

One of the staffers walked by the pool and then stopped abruptly. He had seen something unusual out of the corner of his eye. He doubled back and took a closer look. What he had seen was Mark's body.

The staffer ran to the manager. They came back with the other staffers. They pulled the body from the pool. A lifeguard doing double duty that night as a waiter turned Mark on his stomach and forced liquid from his lungs. By that time, an ambulance had arrived. One of the attendants tried CPR, but to no avail. Mark was gone.

ALL HIS LIFE, he had wanted to be someone other than himself.

43

KAREN

Karen was born in 1950 to Mark and Nora. She was a socially precocious child. When she was a little more than two years old, she was trying to enter conversations at the dinner table. By the time she was three, she was succeeding, on a simple level.

They had developed a standard way to start dinner. When they were all seated at the table, Mark chucked Karen under the chin and asked, "Whadja do today, Sweetie?"

"Umm. Swim. Right mommy?"

"That's right, dear. We went to the Club and you took a swimming lesson." Looking at Mark, "She's doing really well. The teacher thinks she is going to be an excellent swimmer."

"Good for you, Karen. You're a good girl."

EARLY CHILDHOOD WAS CONFUSING. Karen lived in opulence and had everything she needed or wanted.

Except parenting.

Busy with work and social affairs, Mark and Nora delegated parenting to the Club and to domestic help.

BECAUSE OF HER father's erratic behavior, Karen's home-life was grim.

Once she started school, Karen developed a dread of coming home. The first thing she did was check to see if her father's car was there. She was relieved if it wasn't. But she was worried if it was. She had no idea what he might do.

If Karen had friends at the house, Nora watched the driveway. At the first sign of Mark, she ushered the girls into the family room and closed the doors to it.

THINGS CHANGED after her father's passing, when Karen was nine.

Her mother was despondent for a while, crying a lot and keeping to herself. But that passed.

Now they had a decent home-life. They could invite people for dinner. And Karen could have friends over after school.

FROM AN EARLY AGE, Karen spent a lot of time at the Club, until she started school.

At first, it was little more than babysitting and supervised play, with a goal of developing social skills. This included group activities such as birthday parties and holiday celebrations. Karen enjoyed the play and the activities. And she made friends easily, getting to know everyone in her group.

As the children developed, they were introduced to athletics—swimming, tennis, and, several years later, golf.

NORA DROVE Karen to her elementary school during the initial week of her first grade. After that, because Karen seemed comfortable with her transition to school, a maid drove her. Karen never inquired about the change.

During a recess in second grade, she sat down in a part of the playground that was graveled. She noticed a fossil and put it in her pocket. Then she started hunting for fossils. She had no idea of the passage of time. All at once, she was startled to see her teacher standing over her and to see that the other students were gone.

"Karen! What in the world are you doing?"

"Lookin' for these," holding out her hand, with a fossil in it.

Exasperated, "Karen, recess was over twenty minutes ago. Everyone else got back in time. Why didn't you? I was worried sick that something had happened to you. Why did you make me come find you?"

The teacher, like most authority figures, intimidated Karen, who froze with fear at the teacher's anger and could barely speak. The most she could manage was, "Sorry, mam." As she and the teacher walked, in silence, back to the classroom, Karen felt ashamed.

Why am I different from the other students? I wonder if there's something wrong with me?

That afternoon, the teacher wrote a letter to Nora and Mark.

"I write to inform you of a tendency I've observed in Karen. She didn't return to class today when recess was over. I waited about twenty minutes, and then started looking for her. I found her outside, sitting in a graveled area. She didn't notice me until

I spoke to her. She was completely absorbed with looking for fossils.

"I've noticed the same behavior several times in the classroom. If she finds something that interests her, she locks onto it, and it becomes her whole world. She can't seem to think of anything else.

"This letter is not a complaint. I enjoy having Karen in my class. It's obvious that she's not interested in the course work, but she does it and is well behaved. I write because I thought you would be interested in my observations."

DURING FIFTH GRADE, Karen got to know Suzie. Suzie was a little bizarre, but she was outgoing, and that was sufficient for Karen.

One morning, Suzie was talking with animation to a small group of girls. Karen was curious and joined in.

"I've figured out that Ms. Cardwell is a witch masquerading as a teacher. I saw her putting a spell on a little boy. And I saw her teaching a dog how to talk." And she went on with other examples of Ms. Cardwell's escapades as a witch.

The girls were fascinated, some listening with eyes wide open.

Having hooked her group, Suzie announced a strategy.

"We need to search Ms. Cardwell's classroom to see if we can find hard evidence that she's a witch. We can go during lunch. She always eats in the cafeteria with other teachers."

Dramatic pause.

"Any volunteers?"

All the girls, including Karen, excitedly indicated that they wanted in on the foray.

They huddled and decided to go that day.

"Now, we have to carefully make a plan and, once decided upon, carry it out without question. We have to avoid detection. We'll be in trouble if we get caught."

The girls solemnly nodded their agreement.

"We'll gather at the cafeteria door at precisely 12:20. This won't call attention to us. It won't seem unusual. And while there, we can confirm that Ms. Cardwell is in fact eating. See you then."

Karen was excited as she walked to the cafeteria.

Whether or not Ms. Cardwell is a witch, it's an adventure to search her room, which will certainly violate some rule.

Suzie got them into the classroom and shut the door behind her. She gave instructions sotto voce. "I'll take the closet. You guys look in drawers and bookshelves. See if you can find any tools of witchcraft."

They looked, even through the teacher's desk drawers, but found only lesson plans and school supplies.

After rummaging, Suzie pulled a soup ladle from the closet and victoriously displayed it for all to see. Again, sotto voce, "She uses this to pull eyeballs from her cauldron. I've read about that."

The girls twittered excitedly.

"OK, we have our evidence. She's a witch and probably convenes her coven after school in this very classroom." She looked around the room, taking in the feel of it, as if, with her new knowledge, she was seeing it for the first time. "It's been a good operation. Let's carefully put everything back where it was and leave."

Pause.

"One at a time. Let's leave one at a time."

Unbeknownst to Karen, she had suspended her disbelief in Suzie's game long enough to permit a bit of fantasy and whimsy in a life that was beginning to appear, at least during the school year, as tedious and boring; any interruption was welcome.

I don't really think that Ms. Cardwell is a witch. But I enjoyed looking for the evidence anyhow. And it was a group activity, which I liked. I felt like I belonged.

Karen performed adequately with the limited scholastics of elementary school, if she applied herself. But she was inconsistent in doing this. If she was studying, or sitting in a class, she was easily distracted. She was once working diligently, in the library, on a homework assignment, when she noticed some girls playing, and off she went.

Nora told her, "You should do your best. Nobody can ask for more than that."

KAREN ATTENDED Club functions when she was in junior high. The Club's dress code for children of junior-high age was different than that for younger children. Girls wore dresses, rather than pants. So, shopping took on a new meaning for Karen. And she delighted in it. She shopped at the very best stores and bought whatever she wanted. Cost was not a question. She just said, "Nora Riggins," and the clerk wrote up a bill, which Karen gave to Nora.

The Club provided a music and dance activity for seventh graders. Once a month, right after schools let out, the Club had a live DJ, an undergraduate student at Tulsa University, playing popular music in a room set aside for seventh graders. The room was ideal because it had a dance floor.

The first session was sedate at the outset. People just stood around listening to the music and quietly talking. But that didn't last long. Karen grabbed a boy's hand, lead him to the dance floor, and started doing the twist. Karen's example comforted the rest of the children, and they joined in.

After the session ended, a girl asked, "What brought you to start the dancing? I thought about it but was too embarrassed. I admire your courage."

"I don't know about courage. But I do know that if you don't dance, it's no fun. You have to grab the opportunity."

A couple of months later, Karen was talking with a girl at the Club, "I really enjoy the DJ programs. I insist on coming to the Club directly from school. I want to be here as soon as they begin. It's much more fun to listen to music with other kids than alone at home."

For eighth graders, the Club substituted dance lessons for the DJ programs. The point of the lessons was ballroom dancing. It was more decorous for Club functions than the free-form dance styles that were popular.

Nora asked Karen, "What do think of the lessons?"

"They're boring. I can't imagine people dancing like that."

The ninth-grade activity was monthly dances at night. The girls wore party dresses with flats. The boys wore sport jackets and wool slacks.

The Club didn't have special programs for high schoolers. But Karen still spent a lot of time there. She played tennis or swam, usually organizing lunch with people who were doing the same thing. She participated in birthday parties and holiday celebrations. If there were other girls around, Karen suggested going to the bar, where they enjoyed girl talk over Cokes. And she enjoyed other amenities of the Club.

During summers, she often spent whole days at the Club. She could spend hours sitting by the pool, talking to whoever walked by.

Because she was so attractive, she was very popular, both at the Club and in school, especially as she got into the upper grades of high school. She was the envy of the girls and the dream of the boys.

Her academic record in junior and senior high was lack luster. She drifted through her courses, just going through the motions. She easily tired of dealing with what she regarded as the drear of daily life.

In the fall of Karen's junior year in high school, Nora pushed her to decide on a college.

"Aw, mom, I'm not interested in school. I'm not sure I wanna go to college."

"Karen, I'm aware that you've never been interested in school, but sometimes it's wise to do something that's good for you, even though you don't want to. I'll bet that almost all your friends will go to college. Why don't you ask some of them?"

After thinking it over, Karen relented. "Mom, where do you think I should go to college? I have no idea."

Nora obtained catalogs from several out-state women's colleges and shared them with Karen. "Take a look at these and see if any of them interest you."

Karen studied the catalogs and made a decision.

I wanna go to William Woods, in the middle of Missouri.

Nora thought, *there's a male school nearby, so there will be some balance.*

―――――

"THANKS FOR COMING to my office, Karen. I wanna carry out a request that your father made when he signed his will. He asked me to explain your inheritance, when you graduated from high school.

"I'm sure you're aware of some of the assets your father owned. I know you're aware of the company. But you might not know that Nora's house is in his name alone. And you might not be aware of his other assets—stocks and bonds, art, farms, undeveloped land, and miscellaneous other assets."

"You're right. There's a lot I don't know about the assets."

"He left everything to you, in your name alone. So now, the house is in your name alone."

"Wow." Pausing to think. "I'm gonna have to let that sink in." Another pause. "What's the value of the assets?"

"At least five hundred million dollars. Probably more."

"Oh…I'm…I'm…stunned."

"I can see why."

She sighed, heaved her chest, slumped, and starting crying.

"Karen! What's wrong? Are you OK?"

Regaining her composure, "Yes. I'm OK. I'm just worried about managing all this. I don't think I can do it."

"You don't need to. A bank holds title to the assets and manages them. For example, a bank employee is on the board of directors of the company and is quite involved in its affairs. Things can continue this way as long as you want."

That's a relief, Karen thought.

Karen was concerned that her mother would object to the terms of the inheritance, but Nora thought they were fine.

"Do you have a minute, Karen?"

"Sure, mom. What's up?"

"There's something I want to tell you before you leave for college. I've been working up to it, so I'd like to do it now."

"That's fine. But if you've been working up to it, I'm worried that it's serious."

"Well, it's serious to me. I want to apologize for my shortcomings in dealing with your father when you were very young."

"Aw, mom. You don't have to apologize."

"I appreciate that, but I want to. Anyway, I just couldn't stop his drinking. If I spoke to him about it, he got furious. He'd deny drinking excessively and tell me it was none of my business. I was frustrated with him and worried about you."

"What were you worried about?"

"Well, that he'd get angry about something and hurt you."

"Could you see signs of heavy drinking before you were married?"

"No. Not at all. Mark was a perfect gentleman. He was handsome, generous, and charming. Everyone, especially eligible women, liked him."

"How did you guys get together?"

"We were seated next to each other at a Club party. When our eyes first met, it was special. I knew something would happen. And it did. He courted me and then proposed. It was a whirlwind romance. It was like a fairy tale. I was on a pink cloud."

"But then," (taking a deep breath) "but then, soon after we got married, he started drinking more. Over time, he became distant and inattentive. I fell from my pink cloud; I was crushed."

"Was he unfaithful? Was there another woman?"

"No. I don't think so. There was never any evidence of that. His inattention was due to alcohol. Alcohol ruined our sex life."

"That's really awful. It's hard to imagine how you got through it."

"The whole thing was just too much for me. I was puzzled about it and didn't know how to handle it. So, as I could see later, I just withdrew, from everything and everybody, including you. I just wanted to be by myself. I feel really bad about your part of that, and I want to make it up to you, if I can."

"Please don't worry about that, mom. I know you did the best you could.

The worst of it was his death, which I think was suicide. That night. I wish I could forget it, but I can't. Mark was different that night, but I couldn't put my finger on it.

"Late in the evening, he walked out to the veranda. Then a server found his body in the pool.

"Alcohol had ruined him. It turned his life into a living hell. I think he had decided to kill himself at the party, and it was a huge relief to him, knowing that he was about to end it all. I think that's why he seemed different that night.

"And it all happened so quickly that I couldn't absorb it. My mourning lasted for months. My dreams had been shattered. I felt terribly alone."

Sobbing, "Oh, mom. I'm sorry you had to go through all that. But I'm glad you told me about it. I feel I know you and dad better now."

"Thank you, dear. It took all the courage I have to bring it up."

"Well, it's the best talk we've ever had. It's explained things I've wondered about."

"Don't get married until you're out of college. Be somebody."

They looked at each other. Then they hugged.

KAREN QUIT William Woods College in the middle of her first semester. She returned home, to resume living with Nora.

It was an in between time for her. She had left school and had come home to stay with her mother. She had not begun anything else. She was single, had no children, and had no job. She had time on her hands.

She missed the structure that had existed while she was in school. But now, when she awoke every morning, she felt shame at being at loose ends. She felt she should be doing things, like other people do.

She tried getting together with friends. But they were busy. Some were married and had responsibilities at home. Some had children. A couple had jobs in family businesses. Others were away at college.

The difficulty of making arrangements to get together took the fun out of it. It reminded her that she was not doing anything and brought on feelings of guilt.

One morning she decided to go shopping. She had not bought any new clothing for herself for almost a year.

She had a great day. Shopping took her away from her herself. She didn't think at all about her situation. And carrying away from each store her carefully selected purchases made her feel like she was accomplishing things.

So, she tried it again two days later, with the same result.

And gradually it became a habit.

She had never lost her love of shopping, which developed during junior high school. But now she was able to indulge it exponentially. She had her own car and could go whenever and wherever she fancied. And she didn't even think about the money she was spending. It wasn't an issue.

Her favorite form of shopping was buying clothes for herself. She wore things more than once, but not many times. If a party seemed especially important, she always wore a new dress or outfit.

The closet life of most of her clothes was never longer than two seasons. She would regularly clear things out and give them to charity.

———

ONE MORNING AT THE MALL, Karen found a lot of things she liked. So much so that she was overloaded with packages and shopping bags. She was in danger of dropping things, particularly because one of the shopping bags was ripping. And it was starting to rain. On an impulse, she jumped into the first restaurant she encountered. It was large, but full. She couldn't see a single open table. Resigned, she shifted her load slightly and turned to leave.

"Pardon me. But you look like you could use some help."

He was tall, athletic, and good-looking in an indoor sort of way. He was well dressed—wool slacks, jacket, and tie.

"Well, yes, I could use some help."

"Come to my table. I have it by myself. You can set down your packages and catch your breath."

He turned and walked toward what she could see was an empty table. She followed and gratefully set the packages on the table.

"How about I help you with some repacking?"

Without waiting for her response, he removed the contents of the tearing bag and distributed them among the other bags. He even consolidated two of the other bags into one, making it easier for her to carry everything.

"I can't thank you enough. That's a real improvement."

"Here," he said, pulling out the chair opposite his. "Have a seat. Join me for lunch. It's starting to rain pretty hard, and it's lunchtime. I would enjoy the company."

Not wanting to seem rude or ungrateful, she took up his offer. As she did, the waitress arrived and took their orders.

"It might be in order to introduce ourselves. I'm Karen Riggins. I love shopping in this mall, and," glancing at the packages, "as you can see, I've had a busy morning of it. I was heading to my car when my load became unmanageable and the rain started. What about you? What brings you here on a weekday?"

"I'm Mordecai Goldstein. Folks call me 'Mort.' I'm a full-time student at the University, studying petroleum engineering. I'm on break, so I decided to shop for underwear and stuff. But I got hungry before I got to a store."

Karen had to smile.

He opened himself up to me, a stranger, and said a lot with a few words. I recognize the family name. Anyone in Tulsa would. Oil money. And petroleum engineering. It all makes sense. And he made me comfortable about being open with him. I don't feel I have to hide myself from him.

"We have something in common. My family is also in the oil business."

"I assumed that from your name. It's quite a coincidence that we would meet this way."

"Yes, it is. Seems unlikely. But there's one way that we're very different. I have nothing to do with the business. But you, you're learning a lot about it. I admire you for that."

"Are you in school?"

"No. In fact I just left school about a month ago."

"What do you mean, 'you left school'? Are you on a break?"

"No. I quit it. After two months in my freshman year. In the middle of a term."

"Why? That's an unusual thing to do. What drove you to do it?"

"I didn't think I was getting that much out of it, and I was tired of it—the rules and the regimen. It was tiresome."

Their food arrived, and after a pause, they changed to chatter, which they continued through the meal.

She enjoyed their conversation.

He's interesting, a good listener, and non-judgmental. I'm usually on edge when conversing, especially with people I don't know well. *I measure my words, concerned about what others will think of me. But with him, I'm at ease. His open, easy-going, and natural manner of conversing is comforting.*

As they got up to leave, Mort looked at her packages. "Let me help you to your car."

"Thank you. I was beginning to wonder how I'd do it. The rain has stopped, but it's still a lot to carry."

When the packages were stowed, Karen got in the car and lowered her window. "Thanks for everything. It was a pleasant interlude."

"The pleasure was all mine." A pause. "I have a question. I wonder if we could go to dinner this Saturday?"

"Sure. I'd enjoy that." She dug in her purse and removed paper and a pen. "Here's my number. Call me, and we'll make the arrangements."

WHEN MORT PICKED HER UP, she introduced him to her mother.

"It's nice to meet you, Mort. Come on in to the living room. Let's get to know each other a little."

"Thank you. I'd like that. I enjoyed meeting Karen, and I bet I'll feel the same way about her mother."

The three of them had a cordial conversation for the better part of an hour.

"Well, I better let you two get along. Otherwise, you might be too late to be served."

They had a wonderful evening.

THE NEXT MORNING, Nora got to the kitchen just as Karen was sitting down to breakfast.

"You two have a good evening?"

"Oh, yes. I enjoy being with him more than any other man I've known for years."

"That makes this a little harder than I had imagined." Deep breath. "I've thought about this, and I decided to say something to you about Mort."

"What's that, mom?"

"I suggest that you consider not continuing a relationship with him."

Caught off guard, "Why, mom? That really surprises me. He's a wonderful man."

"I can see that. And I agree, he's charming and a nice guy. But, how to say this? You don't fit with each other. You run in different circles. He wouldn't be welcome at the Club. People would be uncomfortable. Both of you would feel this and be affected by it. The result would be your throwing away the friendships you developed before leaving for college."

A pause. "What you're saying, mom—it's because he's Jewish, isn't it?"

"Yes." A pause. "It's not polite to point that out. But that's the problem."

A long silence while Karen thought about Nora's words.

"If it's not polite, how would I explain to him that I can't continue?"

"You'd have to make something up." Thinking. "You might say that you're already in a relationship."

Karen left her breakfast and went to her bedroom, to be alone.

As the day wore on, she ruminated over Nora's words.

Mother has been a member of the Club and a significant part of Tulsa society a long time and knows a lot about both. If she's right, I could end up displeasing a lot of people. I couldn't stand that, especially people who're the backbone of my social life.

This concern intensified with each hour. When it reached the point of being intolerable, she picked up the telephone handset and dialed.

"Hello."

"Mort?"

"Yes, it is. Is this Karen?"

"Yes."

"You sound upset. Can I help with something?"

"I don't think so. I have something to say that I don't want to say."

"What is it?"

"I have a relationship with another guy, so I can't see you again."

A long pause. "I have to say that I'm disappointed. I was looking forward to spending more time with you. But I understand. So, thanks for being straightforward with me."

As she hung up the phone, she felt a loss.

To get some peace of mind, I had to make the call. But I regret doing it, nevertheless.

It was a dilemma that she had had to resolve. But it was a turning point for her.

She sat down heavily on the edge of her bed and stared at her hands, cupped in her lap. Having dealt with the fear that had gripped her, she was able to clear her mind.

I've cut off a possibility that might have enriched my life. My world seems smaller than before. I feel lonely.

I think I'll go shopping.

———

SHE MARRIED TONY REARDON.

He met all the requirements. A third-generation member of the Club, he had the aura to stand out at the club and elsewhere.

He was attractive and charming. He made a good first impression. He was well connected. He had a good job at a leading trust company. He was an outdoorsman. He had served with distinction in Vietnam.

Nora was pleased. "It's a good choice, Karen. He's well liked. He's a bit of an introvert, but he overcomes it. I'm glad you were attracted to him."

"It isn't a matter of attraction, mom. It's a matter of need. I need a husband. How many spinsters do you know at the Club? You have to be married to fit in. And having been raised in the Club, he has the social graces that I need in a husband. What's more, Tony and I have become an item at the Club. I feel I couldn't live with the disapproving looks I'd get if I broke it off.

"Don't get me wrong. While I didn't fall for Tony the way you did for dad, I love him. He's very eligible, and he's pleasant. And I very much enjoy being with him. It was time for me to make a move. So, when he proposed, it was an easy decision."

Nora planned the wedding. Large, elaborate, and formal. The Tulsa newspapers billed it as the most noteworthy wedding in at least ten years. It went without a hitch.

Nora helped Karen purchase a suitable house. Karen engaged an interior decorator to redo the inside. And she bought new furniture for it.

The honeymoon was in France. Tony and Karen stayed in a seaside villa in Marseille.

The couple returned to Tulsa to begin married life, in a redecorated house with new furnishings, two maids to shop, cook, clean, and do laundry, and a lawn crew.

THERE WAS a knock at the door, and the maid brought two well-dressed visitors into Karen's living room. Karen at first greeted them, thinking they were her mother's friends. When they asked for a donation to a charity event, she interrupted them. "I'm not going to buy anything and I'm not going to give anything away." Without more, she ushered them to the door.

Tony, getting up, "You were brusque with them. Was that necessary?"

"Yes. I intended to be rude."

"Why? That's unlike you."

"My grandfather worked his way up from nothing. My family has a tradition of being self-sufficient. I don't want people asking me for a hand-out. I want word to get out that it's a waste of time to come here."

Tony felt like chiding her, but he thought better of it.

I want to save my fights for more important things.

"We should have a front gate," Karen said.

KAREN WAS SOON pregnant with twin boys. The birth went easily, and the boys were given separate rooms. Karen had round-the-clock help, to maximize her rest. In time, she lost her excess weight and regained her pre-pregnancy shape. She was once again a stunning female, although now a mother.

The pediatrician said that the boys would benefit from daily fresh air. Karen bought a double stroller and tried walking them in the morning. But she soon tired of it and asked the maids to take over.

From the outset, the boys were surrounded with expensive toys. Some dangled over the crib. Some were loose in the crib.

When the boys started crawling, new toys were introduced. They were larger and resided on chairs and tables, as well as on the floor.

One of the maids rotated the toys every couple of days. There were so many of them that the closets wouldn't hold the ones out of service. So, the maid stored them in boxes in the garage.

TOYS GAVE Karen a reason to start shopping again. During the final term of her pregnancy, and for six or seven weeks after delivery, she had not been out much.

So, she shopped for toys. She wanted the boys to have plenty of them and only the best. She shopped every week, becoming a fixture in toy departments and toy stores.

A visiting friend asked, "How do you decide what to buy?"

"It's simple. When I see a cute toy that can enrich the boys, I want them to have it."

She quickly learned what was available in stores. And she found that she could shop in catalogs that stores kept on hand for ordering from New York outlets. It was by searching these catalogs that she bought a two-foot-high, leather elephant; a two-and-a-half-foot-high, handmade, scale model of a steam locomotive; and a pair of eighteen-inch suede dolphins.

I'm particularly proud of these purchases.

Her mother asked, "Where did you ever find the locomotive, the elephant, and the dolphins? They are precious."

Karen delighted in answering the question.

At one point, Tony asked, "Do you think you're overdoing it? The boys would probably be just as happy with fewer toys the boxed ones we can give away."

"It's not a matter of happiness, Tony. Infants and toddlers need stimulation to develop properly. I aim all my purchases at that."

I think I'll leave it at that, Tony thought.

She continued to shop for toys.

Tony, it's time we socialized like adults again. I wanna host dinner parties. Not just any dinner parties, but formal ones where men wear jackets and ties and women wear party dresses. And I want significant guests who engage in interesting discussions." Pause. "My goal is parties that guests won't soon forget."

"That sounds like fun. How many guests do you envision?"

"Our dining room table is a limitation. With its leaves, it'll seat ten guests comfortably. I've thought about options, such as renting a larger table, but decided to make the best of what we have".

"I agree with that decision."

Would you take charge of invitations? That would frighten me, but you'd be a natural at it." Watching Tony for a reaction, "I'd like us to have people who are successful, who do interesting things, and who are well thought of."

"Sure. I'd love to. I'll give it a try. Let's see how I do."

Her next step was to meet with a specialty caterer. She explained the type of parties she was planning.

"What can you do for me?"

"That depends on what you want us to do. Do you want us to cook the meal in our kitchen and deliver it for you to serve? Or do you want more help than that?"

"I want you to do everything. Start to finish."

"Let's start at the beginning. We can prepare hor d'oeuvre and place them in your living room. Then, we can answer the door, seat your guests, take drink orders, and serve them. How does that sound for a start?"

"It sounds wonderful."

"Well before the night of the party, we'll develop a menu and get your approval for it. Then we'll shop for the groceries. We'll bring everything to your house and prepare the meal in your kitchen. After the guests are seated at the table, I'll come out of the kitchen and describe the menu. Then we'll serve each course and clear the table before serving the next course. After the meal is finished, we'll clean up the dining room and kitchen, including washing the dishes."

"That's exactly what I want. I want you to do everything. I don't want to do any of it."

"We can do that."

"And, before I forget, on groceries I want only the best. So, you might have to go to more than one store. Is that OK?"

"Yes, of course. Our job will be to deliver exactly what you want."

"Great. Now, how do we get started?"

"You need to call me before you send any invitations. To make sure we are available when you want to have your party. Is that a problem?"

"No. Not at all."

Pause.

"I must warn you that this type of service will be quite expensive. Would you like me to prepare an estimate?"

"No. And don't you worry about cost. Charge what you think is right and I'll pay it without any questions. Cost is not a consideration. Quality is, but not cost. I want these parties to be a huge success."

Karen got what she wanted, and more. The parties were very special. The caterer knew how to make them special.

Before any guests arrived, the hor d'oeuvres were carefully set up. When there was a knock on the door, one of the two women, both wearing maids' uniforms, answered. She got the names of the guests. After the first couple had arrived, the maid would announce the new guests.

The meals were unusual and well prepared. They consisted of four to six courses. And they were served perfectly by the two maids. Guests particularly enjoyed the chef's description before the first course.

After dinner was finished, the guests retired to the living room. They were surprised to find that the room had been cleared of plates, silver, glasses, and remaining hor d'oeuvres. There was hot coffee on a butler cart, along with after-dinner aperitifs.

Tony, without being obvious about it, led the guests into a topic of conversation. At one party—all the guests had taken exotic vacation trips—the topic was leisure travel. At another—all the guests were either elected officers of, or on committees focused on, local government—the topic was a pending bond issue.

The guests loved it. The discussions were enthusiastic and sometimes a bit heated. But Tony was careful to guide people away from unpleasantness—monopolizing the discussion or making personal comments. Guests usually were reluctant to leave, and some parties lasted quite late into the night.

It was not long before Karen's parties and the names of her guests were noted in the Tulsa society column.

"How are your parties going?" Nora asked.

"Thanks for asking, mom. I think they're going very well."

"Good. I was concerned that you'd worry yourself sick over them. Wondering if you had done things perfectly. You've always been such a perfectionist."

"I've avoided the worry by relying on others. Tony has decided on the guests and has invited them. And the caterer has handled the kitchen. He has suggested menus and has seen to the preparation of all the food. He's expensive, but it's worth it for the peace of mind."

After a number of the dinner parties, Tony commented, "The parties are an unparalleled success. They've gained a reputation in social circles, putting you on the social map. People have angled for an invitation, or even asked outright for one. Only one invitation has been turned down. That was only because of an unresolvable conflict."

"Thanks, Tony. You have no idea what that means to me. And I attribute much of the success to you. You've done a wonderful job of selecting people to attend. And also, with the after-dinner discussions. All I can say is 'Thank you. Thank you.'"

She thought, *I'm gratified by the attention and approval that I'm getting. But I don't want to tell Tony this.*

IT WAS MEMORIAL DAY, a day she always dreaded. It reminded her of her father. He had disgraced the family. Everyone in the Club thought it had been suicide, no matter what the coroner had said. And suicide was a social disgrace. She sighed. And to think that he had to do it at the Club! She shuddered.

Oh well. It's taken years, but I'm back in the good graces of society.

ON REALLY HOT, August days, Karen and the boys went to the pool at the Club for a respite from the heat. Karen drank cool lemonades while she watched the boys cavort in the water.

"Tony, I've come to think that our house limits things I want to do. I'd like to have outdoor parties, but we don't have a patio or even a deck. I've spoken to an architect, and we don't have enough lawn to build either a patio or a deck.

"The ideal arrangement would be a large kitchen that opens onto a patio, allowing people to move freely from outside to inside, and vice versa. But we can't enlarge our kitchen, even if we could add a deck or a patio."

Tony reflected. "I suggest that you speak to realtors about a house meeting our needs. How does that sound?"

"It sounds great. I'll start tomorrow."

She met with three agents, each with a different brokerage company. Each had the same observation: "I'm not aware of a house with those features around this area."

Tony responded, "I prefer staying in this area. I don't want to increase my commute."

"I agree. I also wanna stay in this area. But I think we'll have to build. So, I'm going to talk to that architecture firm about what's possible."

"I think that's a good idea. And you might want to make a list of features you want, before you go. Save your time and his."

The next day she was sitting with an architect at a table in his office, with her list in front of her, on the table.

"So, it looks like you know what you want?"

Karen thought for a moment. "I have some ideas. But I'm looking to you for suggestions."

"Good response. Let's hear them."

"So, it looks like you know what you want?"

Karen thought for a moment. "I have some ideas. But I'm looking to you for suggestions."

"Good response. Let's hear them."

"This whole thing is driven by my desire to host outdoor entertainment, sometimes on a large scale, say fifty or more guests. So, size is important. But then, we'll live there, so practicality is a factor.

"For the upstairs, I'm thinking a master bedroom with walk-in closets, two bedrooms with bathrooms for the boys, a room for a nanny, a guest bedroom, and two offices, one for me and one for Tony."

"And the ground floor?"

"I want a grand staircase off the entry hall. Full width to a point halfway up, then splitting into two half-width staircases, one to the left and one to the right. Also, a spacious living room and dining room, a very large kitchen, a library, and a den. And last, but not least, a double door on the back wall of the kitchen that will permit traffic to and from a large patio and swimming pool."

"I think I have a pretty good idea of what you want. I suggest that we prepare some rough, preliminary plans for you to review. If they meet your approval, we can start by locating a suitable lot in your area. Then, we can move to final plans. How does that sound?"

"Wonderful."

"Now, I need the warn you that what we've been discussing will be expensive. Do you have a top price in mind?"

"No. I want the house that I want. Cost is not a consideration."

"Tony! I met with the architect today! He's going to give us some preliminary plans to consider. He was very positive. I think this is going to work out well."

She and he were both pleased with the preliminary plans. She told the architect to proceed with finding a lot. He did, and he found one that would work well.

"Can you buy it for me? I think it would be better for the agent not to know who the real buyer is. Do you agree?"

"Yes, I do. Of course, you will have to approve the price. And we will charge for my time. Is that all right with you?"

"Yes, of course. And my preference is for you to get the best price you can, and I will accept it. Now that I know about the lot, I'm anxious to get it bought before someone else does."

"I understand. And to move things along, I suggest you come in right away to review the final plans."

"That'll be fine. I can't wait to see them. Will tomorrow work?"

The architect met with her the next day to review the final plans. "We'll start with the upstairs. If you have questions or comments, please interrupt me."

She listened with rapt attention. Upstairs, staircase, ground floor, and deck/pool.

"Oh, I like it. I really like it. It's more than I had imagined. I can't wait to see it when it's done. How soon can you start construction?"

"As soon as we buy the land."

Construction took more than a year. Karen spent a lot of time with it. She visited the site at least twice a week. And met with the architect weekly.

"Karen, it's time we discuss finishing. Do you want the house to have a formal feel or a casual feel?"

"Can you give me examples of the difference?

"Yes. Take the living room. If you have ceiling moldings, that's formal. And door and window frames are part of the feel. If the wood is milled, that's formal. And wall coverings make a difference. Wallpaper or textured plaster is formal."

"I definitely want formal. I want the house to be first class, throughout. And I'm going to leave choices to you. I trust your judgment. If you need imported marble for something, get it. I want the house to convey a strong impression of good taste and high quality.

And, as I've said before, cost is not a consideration. Quality is.

The contractor turned over the keys in April. The next day, Karen showed it to her caterer.

"What'd you think? What's the best way to use it for an outdoor party? Maybe fifty guests?"

"To start, people will come in through the front door. We'll have hor d'oeuvre in the living room, so people will see that they're welcome to sit there. The dining room table will have dinner set out buffet style. We'll have two people stationed in the living /dining room area. They can take drink orders and serve hor d'oeuvre.

"For most of the evening, the kitchen will be a walk-through area, to get outdoors and back. Later, desserts and coffee will be set out in the kitchen. That'll be a signal to guests that it's time to wrap up conversations and go home.

"Around 10:00, after setting out desserts and coffee, my crew will do a light clean up and leave. A further signal that it's time to go home. And we'll return the next morning to clean thoroughly and wash dishes."

"We'll want two maids in back to serve beverages and hor d'oeuvres and to remove used dishes and silver. I have only two. Could two of yours fill the outdoor positions? I can get suitable uniforms and give them whatever instruction they need."

"Of course. I'll have two women available.

"Gosh. You think of everything."

"Well, planning is what makes a good party. And I have a lot of experience, which helps. So, how does all that sound?"

"It's exactly what I want. How soon can we start?"

"When do you expect to finish furnishing the house?"

"Two weeks, max."

"Might as well wait until May. That'll be the start of good weather."

"Ok. I'll come see you about menus."

"Around 10:00, after setting out desserts and coffee, my crew will do a light clean up and leave. A further signal that it's time to go home. And we'll return the next morning to clean thoroughly and wash dishes."

"Gosh. You think of everything."

"Well, planning is what makes a good party. And I have a lot of experience, which helps. So, how does all that sound?"

"It's exactly what I want. How soon can we start?"

"When do you expect to finish furnishing the house?"

"Two weeks, max."

"Might as well wait until May. That'll be the start of good weather."

"Ok. I'll come see you about menus.

The next day, Karen showed the house to her decorator.

THE FIRST PARTY was early in May. The weather was nice, and the lawn was perfect. Karen had a full-time lawn crew of three, and they had been doing a good job. The trees were expertly trimmed. Bushes and shrubbery looked sculpted. The grass was the color of an emerald.

The party was a huge success. Tony had invited twenty couples, and everyone enjoyed themselves. Most of them took a moment to talk to Karen, to complement her on the house and the party.

"Tony. Winona was not good with our guests last night. She was arrogant. When someone smiled and thanked her for a drink or an hors d'oeuvre, she didn't respond. She just looked through them, as if they weren't there. I have to let her go. She doesn't fit the job. I should never have hired her in the first place. Indians think they're special, and they're too proud to serve whites."

"That's too bad. She probably made people feel uncomfortable. I agree with your decision. I hope you can be gentle about it."

Based on the first party, Karen and Tony decided to have one each month through September. Their guests were interesting and varied. During that first summer, guests included Oklahoma's Governor, Tulsa's Mayor, business leaders, and clients of Tony's trust company.

NORTHEAST OKLAHOMA HAD BEEN UNUSUALLY dry and hot that August. The newspaper reported that August of 1985 would go down as the hottest and driest August since the Depression.

Aggressive watering was required just to keep lawn and plants alive. So, Karen hired a person to do nothing except water —trees, bushes, flowers, and grass. While other lawns in Tulsa withered and dried, hers was still perfect. It was an emerald oasis in the midst of brown, heat damage.

THE EVENING of the August party was okay. The heat of the day gradually faded, replaced by cooler humidity. The party was well under way, and Karen took a moment to relax in a chair by the pool. She was happily assessing the situation, when one of the guests interrupted her thoughts by sitting down next to her.

"This has to be Karen. I wanted to introduce myself. I'm Bud and I work with Tony. And I want to thank you for this wonderful party. It must have taken a lot of work to prepare for it."

"Oh, not too much."

Looking around. "This is quite a house. I understand that you took charge of the project."

"It wasn't bad. I had a lot of help. But thank you for noticing."

She regarded him—*what was his name again? Bud?* —as a bore, but his wife was from money and socially prominent. They added to the profile of the party.

They talked small talk for a while. Then, suddenly, "How did your grandfather get into the oil business?"

Karen thought, *that was an insensitive question. But I want to be polite.*

"I don't know. I heard a story—something about an Indian and forty acres—but I didn't pay attention to it. It didn't seem important."

Later that evening, when the guests had left, Karen talked to Tony about Bud's question.

"He had the nerve to ask about the source of my money."

"What did you tell him?"

"I wanted to tell him that it was none of his business. But I didn't. I told him the truth—that I don't know. But the whole thing was embarrassing to me. It spoiled the rest of the evening. He probably thinks that I'm a dunce."

"I agree that his question was inappropriate. But Bud's very curious about everything, curious to a fault. Sometimes he's a bit rude. He's probably forgotten the conversation by now. It's too bad that he put you on the spot. But I suggest you let go of it."

IT WAS SEPTEMBER. A beautiful day—cool air, warm sunshine. Karen was alone on the deck. She was reclining on a chaise lounge. She let herself relax for a moment. She started thinking about the coming, off-season dinner parties. *Now that we have a larger dining room and dining room table, we can probably enhance what we were doing in the old house.* She nods off, drifting into sleep. She's dimly aware of this, but she does nothing about it. Her body wants to sleep. *Possibly some music... maybe a string quartet...maybe invite more people...might be too many...have to ask Tony what he...*

44

LUCINDA

Lucinda was born in 1950 to Jake and Ronnie Perryman. She grew up listening to her father's stories about her grandfather, Sam. The accounts of his life fascinated her, especially as she got older and could understand the meaning of facts, not just the facts.

Sam grew up in one political-economic environment. He had to live as an adult in an entirely different political-economic environment. While he was farming, coping with the differences was not very often necessary. But when he had to quit farming, the coping became an every-day occurrence.

He had to struggle, but he never gave up. He did what he could to survive both the Depression and the loss of his farm. His struggles included putting up with discrimination against Indians. He accepted things the way they were and made the best of them.

I want to live that way.

SHE WAS INTRIGUED that Sam had, with his own hands, built the house that she was living in. She would concentrate on something—a closet, a cabinet, a whole room—and wonder how he had been able to design it, nonetheless build it.

The house was old and had never been fancy. It had been improved and occasionally repaired. But it had held up well. And it had served as a home for a lot of family.

There were no photographs of Sam, so she conjured up an image of him. It made it easier to think about him.

In time, she came to conclude that lack of a formal education compounded many of Sam's difficulties. That led to her life's passion—

I want to educate modern Indians.

Lucinda was an observant and thoughtful child. She was born with an ability to listen, to absorb, and to think about what she had learned.

She was reserved and quiet. And she was self-sufficient. She could get along without constant entertainment.

She did well with solitary play. She liked to pretend that she was a teacher, reading to her "students," a mixture of dolls, serving bowls, and whatever else she could find and place in front of her in rows.

She got along well with other children and developed several preschool friends. She enjoyed playing with them, sometimes in her house and sometimes in theirs.

When she started school, she and her friends did homework together and went shopping together. The other girls were often a little silly. While Lucinda enjoyed a little of this, she was more serious than they. She had few silly moments.

As soon as she could walk on her own, she wanted to hold hands while she walked.

When Ronnie took Lucinda with her on errands, she held

Ronnie's hand and looked up at her while they walked. The same thing when Jake took her with him on errands.

But when Jake and Ronnie went together on errands, or to church, she held both their hands, looking up alternately at one and then the other. She felt really special when this happened.

She got a lot of personal attention at home. Before she started school, Ronnie was with her constantly. And once she started, Ronnie drove her to and from school every day. After school, they got dinner ready and cleaned the living room and dining room.

And Jake devoted all his spare time, nights and weekends, to Lucinda. She always enjoyed her time with him.

He told her stories that he had made up just for her.

They took walks together.

He taught her about softball, and spent hours pitching to her so she could practice batting.

He helped her with homework.

He let her participate in house repairs and painting. She couldn't really help until she was older, but he let her think she was helping. She liked to think of herself as "daddy's helper."

When she was at school, she compared Jake's attention to that of other fathers. She didn't find anyone whose father spent nearly as much time with him or her as Jake spent with her.

As an adult, Lucinda thought, *looking back at growing up, I feel that I was very cared about.*

Ronnie began teaching Lucinda to read well before she started school. They checked out books at a county library. Lucinda loved going there. The brightly colored covers attracted her. And she was awed by knowing she could select any book she wanted.

She quickly learned to read independently. And she soon came to like reading.

She did well in elementary school, from the start. She thirsted for knowledge. She was curious about the world. Teachers delighted in her.

At the end of first grade, Lucinda's teacher arranged a meeting with Jake and Ronnie.

"I recommend that Lucinda skip second grade and go directly into third grade. She's reading above grade level and is noticeably ahead of the other first grade students in terms of academic development. I think it'll impede her further development to stay with the same students. And I think she'll progress more naturally and more rapidly by being with older students. I also feel she's mature enough to handle the changes."

Jake and Ronnie discussed it. They tried to weigh the importance of academic development against personal development. They ended up deciding to accept the teacher's recommendation.

Lucinda adjusted to being with older students quickly and without a problem. She did well through the end of elementary school.

In junior high school, she continued to do well academically.

She also increased her socializing skills and her maturity. She played sports with the other students. Sports were an opportunity to learn how to get along with others.

She tried different sports.

I like softball the best.

She played it whenever she could. One of the two organizers of a game always selected her for his or her team, because she was good.

Softball taught her to share and to compromise. It also taught her to be fair and to accept defeat with poise. And it introduced her to teamwork.

Some of the girls liked to get together with three or four others after school, in one of their houses, to listen to music on the radio. Lucinda initially joined the groups when asked.

The girls reacted to what the disc jockey played, either with feigned indifference or with jumping and screams of approval for the songs that they especially liked—"I Want to Hold Your Hand," "Good Vibrations," "My Girl," "You've Lost That Lovin' Feelin'." Not surprisingly, they all liked the same songs.

As they warmed up, they danced, solo, to feel the beat of the music. Lucinda joined in, but she felt self-conscious. And she came to feel dishonest about the whole activity.

I really don't care about the songs. And I think the whole activity is silly.

She soon lost all interest and quit participating.

In high school she did well in all the core subjects and excelled in English.

She also continued to engage in extracurricular activities.

She played softball on the varsity team. The time that Jake had spent teaching her to bat came to fruition. She had the highest batting average on the team.

She tried out for cheerleading and was selected for the squad.

Lucinda didn't have some of the emotional needs that many of the students had. She didn't need attention. She didn't need approval. She enjoyed meeting and being with other students. She also enjoyed her course work.

As Lucinda had advanced through the grades, she became self-conscious about her clothing. She could see that her clothes were not as nice as those of other girls.

They had a limited clothing budget. So, Ronnie taught Lucinda how to cope with it. They found simple, but well-designed, items. They bought subdued colors. And they chose items that Lucinda could mix and match with other clothes in her closet. By dressing in different combinations, it didn't look like she wore the same things every day.

One day, when Jake was at work, Lucinda asked Ronnie how she and Jake had met.

"We met in a Boston hospital." A pause. "Jake had volunteered for the Army and was injured. The Army sent him to Boston to recoup. Jake's mother wanted to visit him, but she was afraid to make the trip alone. So, I went with her."

"That was really nice of you. And you got to meet dad."

"Not only that, but I got to see a soldier present him with a medal for bravery. I was very touched by it."

"What kind of bravery?"

"I don't know. The soldier didn't explain it and Jake has never told me. He doesn't like talking about the war."

Jake's death, when Lucinda was fifteen years old, devastated her. It left a void in her life. She knew she would miss him terribly. She had to talk about it. She left the house, which was filled with Jake and Ronnie's church friends, and went to the house of her best friend, Aimee.

"Aimee, I'm afraid. I don't know how mom and I can live without him."

"Are you concerned about missing his salary?"

"No, no. I guess it sounded like that. But that's not what I meant. We're used to getting along without enough money. What I meant was that the three of us were close as a family. We fit together well. If one of us had a problem, the other two worked as a team to help. With just mom and me, a vital part is missing. I'm not sure how that'll work out."

"I'm sure the two of you will adjust. Your mom is positive and sensible. The two of you will make a new family, closer than before."

"Will you come with me to the funeral?"

"Let me ask my mom."

In a quiet moment late that evening, Lucinda approached Ronnie. "Mom, are you doing okay?"

"Oh, honey," said Ronnie, sobbing.

"I worry about us without Dad. We were such a happy family. How can we ever be happy again without him?"

"Don't you worry. It's a terrible loss, but we'll adjust to it. Think about what he would have wanted. He would have wanted us to be a happy family again, but with two, not three. If we do that, we'll be doing it for him as well as us."

"Thanks, Mom." Lucinda hugged Ronnie. "That's a wonderful way of looking at it. It'll help me."

The funeral was held in a military cemetery. The service was at the gravesite.

Lucinda had requested Aimee, her best friend, to attend.

When Lucinda and Ronnie arrived, the casket was draped with a United States' flag. Their friends sat with them, waiting for the service to begin. At the appointed time, the pastor from their church conducted the service.

On a signal from the pastor, two men in Army dress uniforms approached the casket. They removed the flag and folded it neatly into a tight triangle. One of them approached Ronnie. She nodded toward Lucinda. In response, the soldier approached Lucinda and gave her the flag. He then stepped back and saluted her.

On cue from one of the two men in uniform, a man in civilian dress, except for his garrison cap, stepped forward.

"My name is Don Stephens. Jake and I were in the same squad, and in the same landing craft, on the day that the Allies landed on the Normandy beaches. Once our craft landed, we all got out and grouped together. I was scared to death and I'm sure every one of us felt the same. We walked a little way onto the beach, far enough to be out of the way of other landing craft. Then we hit the ground. We weren't being fired on, although we could hear gunfire from other parts of the beach, but we didn't want to take any chances. We were smaller targets on the ground than standing.

"Our objective was to climb, and take, the hill in front of us.

"Jake was the first to get up. He stood, crouched, and moved up the hill. No one followed him.

"When he was about half of the way up, some tree branches were thrown aside, revealing a machine gun nest with two gunners. There was a vicious burst of gunfire, and Jake went down. We could see that he had taken a bullet in his thigh. He couldn't get up, nonetheless walk.

"The rest of us realized our danger. If we moved, we would be slaughtered. But if we stayed put, it would only be a matter of time before they saw us and started shooting.

"Instead of lying flat until he could get help, Jake raised himself on his elbows. Then he dragged himself, using his elbows and his good leg, further up the hill. He must have been in terrible pain.

"He was completely unselfish. He was looking out for us; he was not looking out for himself.

"When he had gone about sixty feet, he tossed a hand grenade. When it exploded, we could see that he had destroyed the nest and killed both gunners.

"Because of the risk that he took, the squad achieved its objective. Jake had removed the only obstacle.

"And he saved the lives of everyone else in the squad. I think of him daily. And I'm sure the others do too.

"I came here today from Philadelphia, on behalf of the whole squad, to recognize a real hero, to thank him, to let people know that we still remember him, and to say, 'Good bye.'"

Throughout Mr. Stephens' talk, Lucinda clutched the flag tighter and tighter against her chest. She was filled with emotion.

I'm very proud of daddy. I had never heard the story behind his medal for bravery.

When Mr. Stephens stepped back, one of the soldiers came forward with a bugle.

As the last strains of Taps lingered in the moist air, Lucinda felt a chill in her spine. It was a poignant moment that she would never forget. It always reminded her that Jake had answered the call to duty and had served well. In a small way, he had helped the Allies. He had affected the world, at least a small part of it, by improving it.

I want to live as he had lived, helping others when I can, and making a contribution.

Ronnie had forbidden dating before the sophomore year of high school. Other than that, she didn't limit social activities. But she discouraged an excess of them. She felt that Lucinda needed to apply herself to her studies.

The restriction didn't affect Lucinda. It turned out that there was almost no dating in the freshman year at her school.

In the first term of her sophomore year, she had occasional dates. But nothing caught. She found the boys immature, shallow, and not serious about their studies.

In February, Dale, an Indian senior, began paying attention to Lucinda. He sought her out before and after school. If he could, he waited for her outside a classroom and walked her to her next class.

In time, he asked her out on dates. And he taught her to drive.

I like him, Lucinda thought. *He's more mature and interesting than the other boys I've dated. And he aspires to go to college.*

In their school, only a very few students—the ones at the top who received scholarships—were able to continue their educations.

Importantly, he's more interested in me as a person than the other boys have been.

He was spontaneous and natural when he talked to her.

I can see why he's so popular at school.

On their second date, they held hands.

I really enjoyed holding hands.

On their third date, Dale invited her to go with him to the senior prom. She was excited but concerned.

I don't have a nice prom dress, or even a nice party dress. But I'm going to accept anyway. I can find something suitable to wear.

By chance, a church friend of Ronnie's offered to lend Lucinda her daughter's prom dress. She had worn it only once, at the prom two years ago, and she and Lucinda were nearly the same size.

Ronnie made a few alterations and it fit perfectly.

On the night of the prom, Ronnie helped Lucinda dress. When Dale arrived to pick her up, he was surprised at how she looked. He had never seen her dressed up. "Gosh, Lucinda, you look beautiful!"

Ronnie was very pleased.

During the prom, they snuck away to take a walk. After they were out of sight, Dale stopped, let go of Lucinda's hand, and turned toward her. He pulled her closer and softly brushed his cheek against hers. She did nothing to avoid the contact. Mutually understanding the moment, they kissed, at first lightly, and then with more passion.

It was her first kiss.

She confided in Aimee.

"I wasn't expecting it. It was a complete surprise. And it was a thrill, a thrill like nothing I've ever experienced before."

"It sounds like it was made in heaven. I'm happy for both of you."

They started dating regularly, deciding to "go steady."

She attended his commencement, where she got to meet his parents. Out of his hearing, they related that Dale spoke highly of her.

Two weeks into the summer, he spoke of finding work. "I've

spent all my time since graduation looking for work. All I've found is part-time, low-paying jobs. I can't figure out the problem. My school record was good, and teachers have written glowing recommendations."

"Don't give up. You're a good candidate; something is bound to show up."

But nothing did.

In August, he told her, "I've heard of a pipeline company in Austin, Texas that's hiring. I've arranged an interview."

When he returned, he couldn't wait to give her the news. "I got the job! The pay is excellent, and the company will pay the tuition for me to attend a local college."

"Congratulations! You've done well." A pause. "When do you start?"

"Right after Labor Day. But don't worry, we'll keep up with each other, and I'll be home from time to time."

They talked by phone several times. But when he called late in September, things had changed.

"Lucinda, this is difficult. But I've met someone here, and I want to date her."

Silence.

"Don't take this wrong, Lucinda. I like you very much. It's nothing personal. But you're there and I'm here."

Long silence.

"It's okay, Dale." A pause. "I'm disappointed." Pause. "But I understand what you're saying."

"Thank you, Lucinda." A long pause. "I wish you well."

"And I wish you well, Dale."

They never spoke to nor saw each other again.

A few days later, she spoke to Aimee.

"At first, I was disappointed. But now I feel rejection. I wonder if I did something wrong."

"You didn't do anything wrong. Circumstances changed. I

doubt that you want to move to Austin. And he certainly tried to find work here. Now he's locked into Austin. The relationship just hit a dead end."

"I suppose you're right. But I still wish he would change his plans and move back."

"You can't make your happiness depend on what Dale does. Why not let him live his life and start paying attention to how you live yours?"

Ultimately, Lucinda accepted that the relationship was over.

I'm grateful that he was forthright with me.

She continued dating during the last two years of high school, but something was always missing. Her experience with Dale had set a standard for a relationship, and the other boys didn't measure up.

In time, she regained self-confidence and got a better notion of how she stood with classmates.

I enjoy much of my dating. But I doubt that I'm going to form a lasting relationship during high school.

During her senior year, Lucinda turned her attention to arrangements for a teaching career. The arrangements were complicated and time consuming.

The arrangements require travel. I can't ask mother to pay for it. Her financial situation has only gotten worse after daddy's death.

Lucinda convinced her mother to allow her to get a part-time job and to use the car to get to it. The job provided travel money and a little more, which she set aside for incidental expenses at college.

She met with the state board of certification, in Oklahoma City, and worked out a program that would enable her to be certified for teaching after two years of college. The program consisted of required courses and sitting for board examinations after each term.

Then she explored the financial aspects of going to college. She began by creating a budget of what she needed and how to get it.

She chose to attend Oklahoma State University, in Stillwater. She met with its scholarship committee and explained her need for financial help. It subsequently granted her a scholarship based her academic record, her participation in softball, her cheerleading, and her aspiration to teach. It was not a full scholarship, but it was substantial.

She then travelled to Muskogee, to meet with Creek Nation officials who administered a program of financial grants to needy Creek Indians for worthwhile purposes. They were impressed with Lucinda and with her plans to help Indians by teaching. They awarded her an annual grant for each of the two years.

I think I can afford the two years of college that I need. I might have to work part time to supplement my scholarship and my annual grant. But I'm confident I can get a job, if necessary.

To create a cushion, she started working full time right after graduation. Her part time employer was all too happy to have her go full time. He had been very pleased with her work.

Lucinda waited until her eighteenth birthday to tell her mother her career plans. That way she could truthfully say that she was grown and had to start her own life.

She was excited about college. By focusing on her career, she had moved beyond high school, on which she had soured as a result of the Dale affair. She wanted something new.

From an early age, Lucinda had sensed a pall hanging over the family. Something wasn't right. She was too young to understand it. But it was real. It was always there. And she remained aware of it as she grew.

From time to time, something nagged at her.

When I take stock, I realize that it's my recurring sense that all is not well.

When she was in her late teens, she finally came to understand the pall.

It was tight finances. As time went by, daddy's wages had not kept up with inflation. Because he was an Indian, he had been denied promotions and pay increases that less qualified white workers received.

Her parents hadn't fought about money, or even argued about it. But they spent a lot of time talking about it. They felt that they had to be very cautious with their resources. So, they carefully considered every expenditure.

While we were not wealthy, I had a rich heritage. I didn't have to look far to find heroes.

College was everything Lucinda had hoped for, plus some.

The campus is like a large town. It has everything I could possibly need.

She was living in a more densely populated area than she had ever experienced. On Saturdays when there were home football games, the population doubled. The air seemed electric, filled with bands and cheering crowds.

There were activities every day and night.

She felt alive with excitement.

I want to experiment with my new life.

But I'm the first member of my family, going back forever, to attend college. I feel privileged. And I feel the responsibility that goes with privilege. I'm here for a specific purpose, and I need to keep focused on that purpose.

She limited the amount of time that she permitted herself to be distracted by the excitement of her surroundings.

During the week before classes began, there was a mixer for Indian students, and she felt she had to go. There were not many Indians at the school, and they tended to stay together, to avoid feeling isolated. She met and talked to a number of them.

Over the next few months, a number of different Indian students contacted her and invited her to join them for coffee at the student union. Most were men, but some were women. She accepted the invitations, but she didn't initiate follow-ups. She wanted to meet people, but she felt she had very little spare time.

In December, "Alex" called her dorm and left word for her. She returned the call.

"Hi! Thanks for returning my call. Do you remember me? We met at the mixer."

"No. Sorry. I don't intend to be rude, but that was a while ago and I honestly don't remember you."

"How about meeting me in an hour at the union for a cup of coffee? See if that jogs your memory."

"Yeah. I'll do that. But, how'll I know who you are?"

"Not a problem. I know who you are."

When she walked into the union, she recognized him.

Wow! Big, good looking guy.

She extended her hand, and he his, and they shook hands.

"Hi. I'm Alex Walker. What's your name?"

"Lucinda Perryman."

"It's good to meet you."

"I do remember you. Where've you been? I've seen most of the mixer people here and there, but not you."

"I've been playing football. I'm here on a scholarship. And football takes a lot of time. We practice every morning. Plus, the coach has an early evening curfew. We have to check into our dorm every evening—study, then get to sleep early."

Lucinda had not known that Oklahoma State was a major football team in the United States. Nor did she understand the demands that a major football school placed on its players.

"Sounds like you're in jail."

With a chuckle, "Sometimes it feels that way."

"Well, what about now? Are you violating the curfew?"

"No. We didn't make it to a bowl, so we're done for this season. But that's another story."

"What's your program here? What are you aiming for?"

"I want to coach football. Start with a high school team. I hope it's an Indian school. I'd like to help the good athletes get into college. And if I do well, who knows, I might get a college job."

"How long have you been here?"

"Three years. I'm a junior."

A pause.

Pointing to the cafeteria, "I want coffee. You?"

After they had settled at a table, Lucinda asked, "What's your major?"

"A high school coach has to teach. Some of the guys take gut courses like geography. But I'm a history major. I like history, and I figure it'll give me an edge in getting a job. Indian history is interesting."

"And what courses does one take to prepare for coaching?"

"Aside from my major, I'm not too particular. I've been taking a variety of courses in the B.A. program."

A pause.

"But that's enough about me. Now it's your turn. What're you up to and how's it going for you?"

"My goal is to teach in elementary and junior high schools, especially schools attended by Indians. I want to help them improve their lots in life by getting an education. Being educated is everything. That's the only way Indians will elevate themselves."

A pause

"But my family has always been short of money. So, I've worked out a program where I can get certificated with two years of college. It's a tough program. All core subjects with required performance levels. But I think I can make it."

"Wow. You must be smart. Did you do anything but study in high school?"

"I played softball."

"Varsity?"

"Yes."

"What position?"

"Outfield. They wanted me for my bat."

"I guess you had a high average?"

"Yes. Four-year average of .421"

"Wow! That's a great average. How about showing me your swing?"

It was getting late, and Lucinda had a sense that everyone else had left the room, but she wanted to make sure. After confirming this, she stepped up to an imaginary plate, waving a bat over her right shoulder. She fixed her eyes on the pitcher. When the pitcher released, she moved her eyes downward slightly, following the ball. Then she swung the bat smoothly and followed through by completing the full arc, ending with her arms extended and the bat on her left side.

"Beautiful! Good form. Where did you learn that?"

"From my father. He spent hours practicing with me."

"You're lucky that your father spent that much time with you."

"Oh, I agree. I was very fortunate."

She took a minute to think.

Should I tell him about cheerleading? Would that be overdoing it? I don't want to show off, but I want to keep the conversation going.

"But I did more than softball. I was also a cheerleader."

"Show me your moves?"

She checked again, to make sure the room was still empty. She took a stance, hands on hips. She did a couple of kicks, a couple of twirls, and a couple of jumps. Then she pumped her arms, as if she were shaking pom-poms. All this with a winning smile.

Abruptly, she turned around, facing away from Alex. She returned her hands to her hips and looked backwards over her left shoulder, slightly tilting her head. Her expression changed from a winning to a suggestive smile, as she did a bump and grind, swiveling her hips provocatively.

Laughing, as she sat down, "We just did that for fun during practice. We never did it during a game."

Alex loved it. He laughed and pounded on the table with his fists. "After that show, I have to ask you out. Would that be okay?"

"Yes. I'd like that. But I still don't have my own phone. You'll have to call the dorm phone."

"That's no problem. I'll do it."

They got up to leave, and Alex said he wanted to walk her to her dorm, which she said would be fine. They engaged in small talk on the way. When they reached her dorm, they said goodnight and Alex left for his dorm.

When she got to her room, she sat down and reflected on their time together.

I enjoyed it. It was a productive break from studying. He was fun to talk to. He seemed interested in me as a person, not just as a sex object.

I'm surprised at how uninhibited I'd been. Even though I had just met him, I felt comfortable with him. And shedding my usual inhibitions felt good. I let go for a few minutes.

I hope he'll call again.

He did.

And regularly.

They started with the date that he had mentioned at the

student union. That went well, and they both wanted to continue the relationship.

They had a candid discussion about the limited time each had to date. So, they explored other ways to see each other. They both had meal plans, so they took meals together. And they studied together. When their routes converged, they walked together between classes. Between one thing and another, they saw each other daily.

Without discussion, it soon became an exclusive arrangement. Neither dated anyone else.

The sexual aspects of their relationship developed naturally and easily. They both regarded their sexual expression as a shared responsibility. They were able to proceed by respecting each other's signals.

As part of her certification program, Lucinda had to take summer courses. To be together, Alex got a job in Stillwater. So, they continued seeing each other daily.

In her second year, Lucinda went to some home football games, early in the season. She wanted to see Alex play. Having been a cheerleader, she knew something about the game.

I'm impressed with his play. As an offensive end, he is quick, strong, and has good hands.

In November, Alex sustained a knee injury during a football game. It was necessary to stop the game and carry him off the field on a gurney. His roommate called Lucinda that evening and told her what had happened.

She immediately left for the infirmary. On her way there, she thought about Alex and their relationship.

I'm apprehensive. A debilitating injury to a football player can have emotional consequences. Will he change? Will he still care for me? Will our relationship continue?

It's not come to mind before, but I don't want to lose Alex.

He has a lot of good traits, including patience. Like daddy,

Alex is willing to do what's necessary to obtain an objective, one of which was to get to know me.

He cares about people, is gregarious, and is unselfish. He stayed in town last summer so we could be together. He's self-confident, optimistic, and enthusiastic. And he's unflappable.

I haven't observed anything seriously negative about him.

During her visit, her apprehensions vanished. The injury was not that serious, and he was taking it in stride. He would miss the last two games of the season, which he regretted, but that was about the worst of it. In the future, he would have to avoid strenuous activities and sudden twists, turns, and stops, because he could easily reinjure the knee. He would be out of the infirmary in a couple of days and would need to use crutches for a week or so.

He has the same easy-going manner and sense of humor as before. He has not changed at all.

She began thinking about their relationship.

I'm uncertain where it's going.

She called Ronnie. "Mom how do you know when you're in love?"

"Don't worry about it, honey. You'll know when you find him."

It didn't take long for her to decide that she did know.

Lucinda phoned Ronnie in December and said, "Mom, I'm bringing home a friend for Christmas."

"A boy?" said Ronnie, alarmed. "Where will he sleep?"

"He's not a boy. He's a man. He'll sleep on the couch and won't complain. He's Creek."

Early in the New Year, they talked about their relationship. They both felt the same—they wanted to get married. They picked a Saturday right after Alex's graduation. By that time, Lucinda would be finished with her program. To please Ronnie, they decided to get married in Ronnie's and Lucinda's church, with a simple ceremony limited to close friends and family. Neither could afford more.

Alex graduated, on schedule, in the spring. And Lucinda was certificated at the same time. They were both ready to pursue their dreams.

Ronnie invited them to live with her, which both of them thought was great. Neither had a job, nor any savings. They moved in right after the wedding.

Alex got a real break when he started looking for a job. The Vietnam War was raging, and the football coach of a nearby high school had been drafted in June, leaving the school district adrift.

During Alex's interview, someone asked about his draft status. By chance, he had a ready answer. When he was discharged from the infirmary, the doctors gave him a copy of his medical records and told him that he should promptly take them to his draft board. He did this and was classified 4-F, exempt because not qualified for military service by reason of vulnerability to knee injury.

All the students in the district were Indians. Because of the district's proximity to Tulsa, and because of the importance of football in Oklahoma, district officials thought that a competitive football team was important for all its students, especially those in high school. Practice would start in August, and a new coach would need lead time to evaluate staff, facilities, and players and to create a plan for the season. So, the officials felt a time bind. Because salaries were fixed, they decided to take the unusual step of offering a signing bonus for a new coach.

Alex liked the sound of the position. It would present challenges, which is what he had hoped for.

The next day, he got a call telling him that the job was his. He drove to the school to sign his contract and pick up his bonus.

He spoke to Lucinda as soon as he got home. "I want to use the bonus to buy a washer and dryer."

"Oh, Alex. That's sweet of you. But don't you think we should put it in a savings account for contingencies?"

"No. I don't. In the short time that we've been here, I've observed your mother doing laundry. It's really a chore, and I think she deserves better. And besides, she's not charging us to live here. I'd like for us to contribute something early on."

"Well, it's your bonus, so I'll defer to you. And, to be clear, I think it's a wonderful gift."

The next day, they went shopping. They bought two top-line units. The price included plumbing, electrical connections, and installation.

A week or so later, Lucinda heard that a small, rural school in their area was hiring a teacher. All the students were Indian. It was exactly what she wanted. She called and made an appointment with the principal.

She borrowed Ronnie's car to get to the appointment. She was not used to driving and didn't notice that the car was low on gas. So, she ran out on the way. She walked the rest of the way to the school.

She apologized to the principal. "I'm really sorry I'm late. I ran out of gas and had to walk the rest of the way. I hope I haven't inconvenienced you."

He just looked at her for a minute or so without saying a word. He finally broke the silence. "Say no more. You're hired. I need teachers with your dedication. I'll drive you to your car and give you some gas. On the way, you can tell me what subjects you can teach."

She started in August, a month before school began. The principal wanted her to teach English, which pleased her. She prepared lesson plans and discussed them with him. One objective was to instill the value of education in the students. Her plans included regularly providing evidence of that value, using examples of specific people. Some were known to the students—

professional athletes, musicians, businesspeople, and bankers; most lived in the immediate area.

It didn't take long to establish herself. She was an immediate success with other teachers. Her students liked her and took her course seriously. Teachers of other subjects could see improvements in most of her students.

Parents of the students were also pleased with her. They could see improvements in how their children wrote and spoke English. And they observed an increased interest in school, along with higher grades.

She worked hard. She stayed late at school to help students who needed special attention. And she often stayed up late at home, adjusting a lesson plan or working on a way to help a particular student.

Lucinda and Alex needed separate vehicles for their jobs. He had a pickup truck, which was fine for commuting. They took out a loan, to buy a used car for her.

I don't like debt. But I see the necessity of it in this case.

With both of them working, and no children, and no rent, they were able to make the car payments and to save a little money.

I'm relieved. Growing up, we never had any savings.

But things changed when Lucinda had their first child. She had to take a maternity leave from work, which was unpaid. And there were extra expenses for doctors, diapers, and baby foods.

After her maternity leave, she returned to her job. So, they again had two incomes, and they were again able to save some money, Lucinda's big priority.

Things changed again when she had their second child. After the maternity leave for this child, she again returned to work.

Although finances got tighter, Lucinda and Alex were thrilled with their two boys. Ronnie took care of them during working hours. But Lucinda and Alex spent all their spare time with them.

They lost Ronnie when the youngest child was three. She had not been feeling well and was diagnosed with cancer. She lived only six months after the diagnosis.

Alex's parents took over childcare. They didn't live nearby, and they didn't like driving. So, Lucinda and Alex worked out a system to take the boys to their house and to pick them up.

Between Ronnie's medical expenses, occasionally helping Alex's parents, paying off the car loan, and the costs of a family, Lucinda and Alex had used their meager nest egg and were not able to get back to saving. They were living hand to mouth.

She would, from time to time, wonder: *Why do some people have so much and others so little?*

It was Memorial Day. In a tradition established by Lucinda's mother, the whole family visited the National Cemetery where Jake was buried. They paid homage to a real hero. By going every year, she hoped the boys would feel the pride in their grandfather that she still felt.

As part of the tradition, while the family stood around Jake's headstone, Lucinda repeated the story that the old soldier had told at the funeral. The story still gave her goose bumps, just in the telling.

Like any Indian in eastern Oklahoma, Alex grew up with sensitivity to discrimination and racial animosity.

Once he got out of school and into the workplace, Alex heard about and witnessed animosity and discrimination toward Indians that he had not heard about or seen before. He personally encountered it when finding summer work. But he always found something.

He thought, *I would have gotten a better summer job with better pay if I were white. I'm beginning feel akin to the Native American-rights movement.*

But he didn't participate in the movement. No marches; no sit-ins; no demonstrations; no protests. He kept his feelings to himself, except for Lucinda and a few trusted friends.

When he saw racial injustice, it rankled him. But he always kept his cool. Except one Friday night at a football game.

He had been coaching for ten years. His team, the "Tigers," kept doing better every year. And his players got better. A number of them had signed letters of intent with Division 1-A colleges. Two of them were in the NFL.

For several years, the Tigers' record had caught the attention of sports reporters for the Tulsa newspapers. They attended games and even brought photographers with them. Because of this publicity, Tulsa teams wanted to play the Tigers.

A match was arranged between an all-white Tulsa team and Alex's team, to be played in the Tigers' stadium. The game was particularly physical and close. At half time, both teams were scoreless.

A jeering white fan of the Tulsa team was sitting close to the field, behind the Tigers' bench. He had been shouting racial slurs all night: "The only good Indian is a dead Indian." "Push those squaw hoppers back." "Injuns are too dumb to play football." "Custer should have won." "Stop those Redskins." "Back to your teepees." "Get that buck quarterback." From time to time, he would let out a war whoop.

Alex walked up and down the bench, telling the players not to give the guy the satisfaction of acting bothered. "Mind the game," he said. But he could see that they were upset by him.

At half time, Alex watched his team walk off the field, going to the locker room. He could see they were demoralized.

Then he jumped the guardrail and walked into the stands. He got close to the fan, looked him in the eye, and quietly asked him to stop what he was doing. Without warning, the guy clenched his right fist and threw a cross punch. Alex was faster than the fan, evading the blow by leaning to his right. To deflect a possible, follow up punch, Alex bent his right arm at the elbow, to a right angle, raised it so that his clenched hand protected his face, and twisted slightly so that he was directly facing the fan.

At that moment, a camera flashed, temporarily blinding Alex. Almost simultaneously, two policemen arrived and separated him and the fan.

Two more policemen arrived. After speaking to the first two, they arrested both men and took them to the county police station. The first two officers arrived at the station a bit later, after interviewing several bystanders. They reported a consistent account that Alex had approached the fan in a threatening way, but that the fan had thrown the only punch. The police decided it was not worth their time to charge either. So, they released them and drove them back to the stadium.

Alex arrived home at 2:00 am.

I'm relieved to see him, but also angry.

"Where have you been? I was worried sick. The game wasn't broadcast, and there was no one I could call. I thought of driving to the stadium, but the boys were fast asleep."

"I'm sorry about this. There was an incident involving a fan. He had upset the team by yelling racial slurs all during the first half. I tried to let it go, but I just couldn't. At half time, I went into the stands and asked him nicely to stop. But he got angry and wanted to fight about it. He tried to hit me, but I avoided his fist. Then two policemen pulled us apart. The police didn't want to argue about it, so they arrested both of us and took us to the station. We weren't there long before they released us and drove us back to the stadium. I'm sorry it happened, but that's the story."

"Oh, Alex." Hugging him and crying, "What a terrible evening." A pause. "I'm glad you're okay. Let's get some sleep."

The phone rang about 4:30 a.m. Lucinda answered.

Reporters wanted Alex's name, his background, his tribe, a statement about what happened. She didn't respond.

The phone kept ringing, so she finally unplugged it.

Tulsa's Saturday-morning paper reported the incident on the first page of its sports section. It included the photograph that was taken just as the police arrived at the scene. Unfortunately for Alex, the angle of the camera made it appear that he was about to strike the fan.

The phone rang at 9:00 that morning. The principal asked for Alex.

"Alex, I'm sorry to call on Saturday, but this is an emergency. The school board members want to meet with you in an hour."

"What's going on?"

"Have you seen today's paper?"

"No, sir."

"It gives prominent coverage to the incident at the game last night. The article, plus the photograph, is bad publicity for the school district—and for Indians. We need to dispel it the best we can with a press release. But we can't write it without your help. None of us were there. We need your side of the story."

Lucinda had listened to Alex's side of the call, and her expression became perplexed as she tried to make sense of it. Alex was upset, but he kept his composure. "The principal wants me at the school in an hour for a meeting to discuss last night's incident. He said the morning paper had a story about it that's unfavorable to the district. Members of the school board want my help preparing a press release. That's all I know for now. I'll fill you in more when I return."

The school's official press release pointed out that the reporter had not interviewed either the fan or Alex and, hence, the story did not describe the fan's provocation and use of profanity. So, the release incorporated Alex's version of the incident.

Two board members had asked him for an apology, but he didn't think an apology was appropriate, given the fan's behavior. They settled on a statement that he regretted the incident, but that he thought racial slurs were inappropriate anywhere, particularly at a high school football game.

The final paragraph stated that the board had placed Alex on probation for a year and that it had cut his salary through the end of the current school year. This was the board's decision on the question of an appropriate sanction. They debated whether to include it in the press release, and decided to do it, to appease whites in Tulsa.

On his way home, Alex couldn't stop ruminating over the events of the night before and of the morning.

I'm anxious about explaining the sanction to Lucinda and concerned about our finances.

After telling her about the meeting and the press release, he apologized. "I feel terrible about this. I'm upset with myself for losing my cool last night." A pause. "I wish I could do something to undo it, but obviously I can't."

"Oh, Alex." Hugging and kissing him. "Don't be so hard on yourself. I'm actually glad that you stood up for your team. More than that, you were standing up for all Indians. It's just bad luck that the police were nearby, and that the reporter took the picture."

"Thanks for that." Choking up, "That really means a lot to me." Long pause, while letting his emotions pass. "But what about the salary cut? Our cushion is small. I'm worried about how we'll make it."

"Well, don't worry, please. That'll just make things worse." Then, suppressing her own fear, "Things have a way of working out."

When Alex walked into his history classroom on Monday morning, the entire class stood up and applauded. They continued for over two minutes, which was about the amount of time it took Alex to get over being choked up from his gratitude over their affirmation.

There was a reprise when Alex entered the locker room that afternoon for football practice. The entire team, the assistant coach, and the field staff stood and applauded. Again, Alex had to deal with being choked up.

And when he told Lucinda that evening, she cried and thanked God, and he choked up again.

It was late in July.

Alex was at work. He had gotten a good summer job.

It had been rough financially getting through the last eight months on his reduced salary. They had cut back, and they had borrowed. And parents of Lucinda's students took up a collection for them. It was not large, but it meant a lot to Alex and Lucinda. And it helped them make it through.

With his summer job, they were able to loosen up a little on expenditures and still pay debts. The debts would soon be fully paid. More than that, they would go into the fall with some savings. Not a lot; but something.

The weather that summer had been almost intolerable—no rain at all and scorching temperatures—unusual for northeast Oklahoma. The view from their front windows was depressing. One could see the dirt road that ran by their front yard, dry as a bone. And one could see the sun-scorched remains of vegetation —dead grass, withered bushes, faded and dusty tree leaves. Even structures held a layer of dust. Everything outside was a dull brown.

The house was more than 100 years old and built entirely of wood. The exterior had been painted regularly but didn't look that way because of the wood's weathering. Surrounding a small patch of grass, dead but mowed for the children, were wild grass and weeds, overgrown bushes, and trees that had not been cared for.

Lucinda was checking on the boys, who were playing outdoors in the intense heat, kicking a can, raising puffs of dust. She had thought about taking the boys someplace cooler where they could play. But she hadn't come up with anything. Swimming pools were closed, to conserve water. And playgrounds were just as hot and dry as their front yard.

A truck happened by, spewing a cloud of dust, which enveloped the boys. Combined with perspiration, the cloud would create dust freckles on their exposed skin and a crust of dirt on their shirts and shorts. This was nothing new. She had been dealing with it all summer.

JUST TODAY'S load of dirty laundry.

If we were rich, and Alex had a desk job, people would applaud what he did as a statement of principle. But we're not rich, and Alex doesn't have a desk job, so people regard what he did as punishable.

What gets me is the school board's inconsistency. First, they give him a signing bonus and tell him to improve the team. Then he improves the team to the point that a white team from Tulsa wants to play against it. Then the board reduces his salary because he takes a stand against overt discrimination. It seems so unfair.

Poor Daddy. He was determined to have running hot and cold water in our house. He got it done, but it put him in a hole that he never climbed out of. That was the cause of our financial anxieties. But God bless him. It made life so much more pleasant. Instead of having to pump water from the well, heat it, pour it into the tub, and sponge the dirt off the boys, I can tell them to get into the shower and to remove the caked dust.

Thank goodness we own our house and it is debt free.

And thanks be to Alex for insisting that we buy a washer and dryer with his signing bonus. I would hate to wash everything by hand, like mom had to do. She used to work so hard to keep our clothes clean.

Still, it's so monotonous. The same thing, every day. Alex gets home from work. Dirty clothes. The boys finally come inside. Filthy clothes. The family sleeps in a hot house—open windows do little to mitigate 104°. Sweaty nightwear. It's the dailiness of it that gets me down.

IT IS LABOR DAY. The summer heat has broken. Lucinda is taking a breather—reclining on a couch.

Her mind wanders.

When things are tough, it always comes down to money, or the lack of it. We can't seem to save enough to ride out the rough times on our own. We have to ask for help.

It's been that way all my life. You can't get ahead just by working. You have to have a lucky break somewhere along the line.

I was fortunate to have the parents that I had. They often had to make do, because they never had enough money. But I still had a great childhood, and during the difficult times of my late teens, Ronnie was wonderful.

Her mind wanders further.

She still remembers the night of the prom. When she thinks about it, which she is doing now, she aches for something. But it isn't for Dale. Its for that first kiss, the excitement of it. She knows she can never again have it, but she longs for it nonetheless.

She starts feeling low. And then she remembers Ronnie's words, often repeated, "Don't dwell on unpleasantness. Instead, help others by improving something. You'll feel better, and you'll be doing God's work."

She gets up and washes the living room curtains.

ALL IS SAID AND DONE: THE 40 ACRES REMAIN

It's 1985. Mid-summer, early in the afternoon. The sun bore down, oppressively. The air is hot and humid. There's no breeze. It's like being in an oven. A person would want to sit indoors, in front of a fan, drinking a large glass of iced tea.

There are four oil wells on the tract. The pumping units ceaselessly go up and down, up and down, lifting the oil to the surface. There is a cyclical squeak, signaling metal rubbing on metal, on one of the units.

There is the sound of a straining electric motor, lifting the heavy down-hole equipment that pushes the oil upward. It is followed by the sound of a racing motor, letting the equipment back down the hole, the weight of the equipment increasing the revolutions per minute.

A bronze plaque is welded to one of the pumping units:

Perryman #1
NW NW 33-18-7
Lew Riggins' first well • Drilled 1913 • Reworked 1968
Riggins Oil Company

There's a rough, dirt road leading to the tract. A battered and worn pickup truck approaches. The driver is the pumper. His clothing is filthy and drenched with sweat.

The pumper, whose sole responsibility is to keep the wells producing, drives onto the tract on his daily routine. He drives to each well and checks it. He finds the source of the squeak and summarily silences it with a single squirt of oil from a long-necked can.

He drives off, only to return tomorrow.

A lone crow flies onto the 40 acres, alights on the limb of a dead tree, surveys the scene, cocks his head, and cries a "caw" of disgust. Then he flies off.

All that stays behind are the four wells, their pumping units in ceaseless motion, producing the straining/racing sounds of the motors.

Sam had been right. There are no cattle on the 40 acres.

LOOKING BACK

A frontier is never a place; it is a time and a way of life. Frontiers pass, but they endure in their people.

CPSIA information can be obtained
at www.ICGtesting.com
Printed in the USA
LVHW081545090821
694908LV00015B/567